How Does Government Listen to Scientists?

Claire Craig

How Does
Government Listen
to Scientists?

palgrave
macmillan

Claire Craig
Royal Society
London, UK

ISBN 978-3-319-96085-2   ISBN 978-3-319-96086-9   (eBook)
https://doi.org/10.1007/978-3-319-96086-9

Library of Congress Control Number: 2018950338

Cover illustration: oasis15 /Getty Images

Printed on acid-free paper

This Palgrave Macmillan imprint is published by the registered company Springer Nature
Switzerland AG
The registered company address is: Gewerbestrasse 11, 6330 Cham, Switzerland

To Dave, John and Mark; and also to all the women in all the multiverses who have already been UK Government Chief Scientific Advisers and all the ones in this universe who will be soon.
With thanks and love to my parents, Yvonne and Richard, to Sukey, Helen and Cathy, and to Chris beyond words.

PREFACE

This book emerges from a passion for science and for democratic government, and from respect for scientists, policy-makers and the usually unnamed people who work closely with them to enable them to do what they do best. It is intended for a reader interested in any of those roles.

It aims to do two things. The first is to help spread the knowledge of practice and theory that is well established in many places but still patchy in others, even more consistently across the landscapes. For example, it should no longer to be possible for the Minister to be surprised when they find the scientist to be comprehensible and interested. Nor should it any longer be possible for the scientist to think that science alone will determine the answer to a policy question. Neither the Minister nor the scientist should, knowingly or inadvertently, allow disputes about narrow points of science to act as lightning rods to distract society from dealing with tougher issues that are less comfortable to debate. The growing body of practical lore on science advice can also be more clearly linked to the various theoretical frameworks in order to help good practice spread more rapidly.

The second aim is to build on that established knowledge by drawing together insights from more sectors and disciplines than are typically included. The practice of providing scientific advice to government has had time to mature; practitioners and academics can reflect on experience of the evolution of evidence and public reasoning in areas from climate change and pandemics, to GM crops and Artificial Intelligence.

It is also possible now to consider the relationships between science and government in the context of futures thinking, narrative studies, public engagement and the behavioural sciences. We are on the cusp of being able to model much more of the world: to create better computational and other models of our social and physical systems. Digital humanities and scholarly insights may, like the introduction of the telescope or microscope, enable us to see previously invisible underpinnings to our familiar worlds such as the ways narratives affect collective anticipations about the future and decision-making. Models and narratives can both be seductive and society needs to have the capacity to reflect on how it is using them in the process of public reasoning.

If, as is the case, every academic discipline is potentially relevant, every area of policy potentially at stake and all knowledge is contingent and uncertain, the discussion can rapidly spiral out to light years of physical space, millennia of historical time and the interconnectedness of the global population and planet. The book retains its focus through the discipline of embodiment. In the end, a human being has to make a decision. In government, the decision-maker usually has to account publicly for their decision. That forces a confrontation with the evidence and with the realities of making a decision about the future now, when observational evidence can only be about the past. It is a White Queen moment, when life is lived backwards as in Alice's Looking Glass world in which the pain (or the pleasure) of making the decision is felt today, even though the real-world outcomes will not happen until later.

Ultimately therefore the book is both about knowledge and about people. It is particularly concerned with specific decision-making by an individual in power, typically a Minister or a Mayor. Alongside the conceptual frameworks it considers what has sometimes been described as the craft skill of the practice of science in government, which depends on personal relationships, empathy and practical detail, as well as curiosity, open-mindedness and rigour.

The interface of science and government can be a lonely place to inhabit and thriving in it requires a willingness to be wrong, indeed to be wrong in many ways. Errors start with the slight loss of accuracy in describing a deep disciplinary concept which is essential to be able to communicate it to a wider audience, to the inevitably imperfect framing of a complex system or wrong judgement of the best moment to consider an important choice. The motivation is usually that being wrong in all these ways is still better than not to have attempted to bring science to bear in the first place. It is in this spirit that this project is undertaken and in the knowledge that in this context, as a former Chair of the UK's Climate Change Committee once said, "any statement that is perfectly true is not useful, and any statement that is useful is not perfectly true" (Turner, 2013). In this spirit, too, the text includes references selected as starting points to further exploration, rather than comprehensive accounts of all of the most relevant literature.

The centre of gravity of the project, like its metaphors, is rooted in the natural and physical sciences. The text refers to science throughout. However, the starting point for considering what forms of academic insight may be relevant to any significant policy question is always that they all are. The natural language of policy-makers and, at least in the UK, their education and training typically appear to share more with the social sciences and arts and humanities than with other forms of knowledge. The tendency to take such links for granted may be one of the reasons there has been less theoretical examination of the ways such forms of knowledge affect policy outcomes.

It is particularly important to take stock of what we know now when it may be that we are, at least in the West, at some inflection point in the accepted roles of scientific knowledge and values, elites and the distribution of power. Yet while public debate asks what it might be to be beyond that inflection point in an area of the graph that is post-truth, post-expert, post-elite, post-normal, the figures still show that, in the UK, public trust in scientists to be scientists (whatever that means) is not falling.

Discussion about science in public life is ultimately nothing to do with CP Snow's two cultures of arts and science. There are multiple academic cultures and, if we are looking for binary distinctions in the twenty-first century, then they are probably rationality (or cognition) and sentiment (or emotion). The challenge is to enable both to play well-founded parts in public reasoning and decision-making.

London, UK   Claire Craig

Reference

Turner, A. (Performer). (2013, February 27). *Distinguished Lecture, Public policy and the science base: Successes and failures.* Magdalene College, Cambridge, UK.

CONTENTS

CONTENTS

ABBREVIATIONS

COBR Cabinet Office Briefing Rooms
GCSA Government Chief Scientific Advisor
GM   Genetically Modified
HFEA Human Fertilisation and Embryology Authority
IPCC Intergovernmental Panel on Climate Change
NRR   National Risk Register
SAGE Scientific Advisory Group for Emergencies
SF   Speculative Fiction
TB   Tuberculosis

CHAPTER 1

How to Create the Conditions Where Science Can Help

AN ILLUSTRATION: SCIENCE ADVICE DURING THE FUKUSHIMA EMERGENCY

The date is 2011, the Great East Japan Earthquake and tsunami have struck and more than 15,000 people will die. Meanwhile, the Fukushima Daiichi nuclear plant is in a critical condition. The outcomes will alter, amongst other things, the future of energy policy in Europe. The question from the UK Prime Minister to the Government Chief Scientific Adviser (GCSA) is whether British nationals should be advised to stay in Japan, or to leave (UK Government, 2012).

The GCSA, Sir John Beddington, invoked the pre-arranged mechanism characterised as the Scientific Advisory Group in Emergencies (SAGE) and over a period of about 48 hours gathered scientists from disciplines that included nuclear science, engineering, meteorology and medicine. They reviewed the available evidence and advised the Cabinet Office Briefing Rooms (COBR) Emergency Committee on the balance of risks to people staying or going.

Based on SAGE's advice about the risks the Prime Minister concluded there was no need to evacuate British Citizens outside the exclusion zone recommended by the Japanese government. Some national governments took a different view, and advised their citizens to leave. Sir John meanwhile carried out telephone conferences with open question and answer sessions at Embassy locations, to which Japanese officials were also invited.

© The Author(s) 2019
C. Craig, *How Does Government Listen to Scientists?*,
https://doi.org/10.1007/978-3-319-96086-9_1

The advice given during the Fukushima emergency shows in acute form some of the characteristics of the provision of science advice in policy more generally. From a practitioner's point of view perhaps the most important are first, that the science does not determine the outcome; and second, that any significant policy or political question requires insights from multiple disciplines. In an emergency, the adviser will need not only rapidly to decide which disciplines are most relevant but also who to bring in to represent them. Urgent advice depends strongly on knowledge embedded in scientists. In that case they must not only be good scientists, but also willing to work flexibly and communicate succinctly. They need to be willing to use their judgement based on the immediate available information about the particular incident, and they need to be tolerant of other disciplines' mental models, even notions of what constitutes good science or scholarship.

A large part of the discussions amongst the experts during the emergency was about how to define and describe the level of risk. Remember that there was little time to refine or define, and the important point was communicating specialised information about risk and degrees of uncertainty to non-scientists such as Ministers, who would in turn have to defend their decisions to the public and in the media. Chapter 2 explores engaging with risk.

While the direct death toll from the Fukushima emergency is low, the wider effects are highly significant. One author states that the Fukushima incident is "extremely unlikely to result in a single death" directly from radioactivity (Thomas, 2011), as does the IAEA's report (International Atomic Energy Agency, 2015). But concern about the direct effects of radiation may deflect attention from the huge systemic effects such as human displacement and the effects on mental health. Chapter 3 explores ways for thinking about systemic effects and plausible futures from the perspective of decision-makers.

At the time there were also reports that some in the West, applying bounded rationality to make sense of the extensive news coverage, may have associated the high death toll with the nuclear emergency rather than the earthquake and tsunami. Sir John's willingness to discuss the situation publicly when it was still extremely uncertain opened up opportunities for engagement with the changing levels of knowledge and confidence. Chapter 4 explores modes of public engagement and the use of different lenses in public reasoning.

The experts grappled hard with the challenge of summarising degrees of certainty and confidence in their evidence. They settled on defining what they called the Reasonable Worst Case scenario, and outlining the risks associated with that. At the moment of Ministerial judgement, cognitive discussions about what risk such a phrase might represent take their place in a wider context of experience, professional relationships and the emotional dynamics of human decision-making. The Minister is really asking the scientist the very human question: "Are you sure? Are you *sure* you're sure?" As Chap. 5 explores, curating these moments is at the heart of the craft skill of enabling policy-makers and scientists to get the best from each other.

FRAMING THE MOMENT

Whether looking at the near or the very long term, the first challenge for the decision-maker and the Adviser is to agree on how to frame the question and the system. Fukushima is an example of what Roger Pielke Jr. calls "Tornado politics" (Pielke Jr., 2007). Decision-makers know they want scientific advice. A relatively small number of people are directly involved in the process of making the decision. Although under scrutiny, the values that inform it are not much contested at the time.

Pielke contrasts "Tornado politics" with the "Abortion politics" that apply when the science is relatively well established and understood, and political resolution depends primarily on judgements about values.

There is perhaps a third category, that of Ecstasy politics. Here there is not yet a reasonable consensus about which science is most relevant, as in Tornado politics, nor are there established value-based or political framings, as in Abortion politics. This category takes its name from the dispute in the UK that followed advice from the Advisory Council on the Misuse of Drugs to the Home Secretary. In line with its statutory duties, the Council considered the harms and benefits of the drug Ecstasy. It came to the view that those harms were insufficient to justify it being classified and controlled under the relevant legislation.

The political context in the UK, as in many other countries, includes deep-seated moral and political viewpoints about the significance of, and ways to manage harm from, drugs in general. The government announced that it would classify the drug. In protest at what he considered to be the government ignoring the advice, a leading expert, Professor David

Nutt, resigned from the Council. He vividly compared the harms from Ecstasy to those of the risky pursuit of horse-riding (Nutt, 2009).

This very public dispute, although possibly damaging to the perceptions both of scientists and of politicians to each other at the time, prompted reflection that led to greater clarity about what should be expected of both. The Government Chief Scientific Adviser led the development of principles that, to paraphrase, pointed to the importance of scientists providing the best evidence; of Ministers listening to it; and of Ministers explaining the bases of their decisions. It equally pointed to the importance of scientists respecting the democratic authority of Ministers in making those decisions, including the requirements on them to take account of factors additional to the scientific evidence (UK Government, 2010).

For those concerned with the proper conduct of science in government the counter-example to the Ecstasy debate was the sequence of events around the government's decision in a similar instance, that of the drug, khat. The Council advised that khat was not sufficiently harmful to justify classification. In line with the new guidelines, the Home Secretary of the day met the Council's Chair and demonstrated that she had understood the basis of their advice. When she nevertheless decided to classify the drug, she wrote to him explaining the other factors that informed her decision, such as the objective of supporting international co-operation between police forces (May, 2013). This form of public reasoning better satisfied those who wished to ensure the scientific evidence was heard, while acknowledging the wider range of factors that inform political decisions in a democracy.

From the perspective of providing evidence to inform the decision, Ecstasy-style disputes are often implicitly disputes about the definitions of the system. Disputes about the definition of the system affect what science is relevant, and therefore also which stakeholder voices have a say. Taking khat as an example, the scientific system might be the human body and the drug's effect on that, the economic and social effects on the user's family, on their community, on national health or security budgets and objectives or, as in this case, on international systems of policing whose objectives go far beyond policing khat.

In Tornado politics, those dealing with the crisis will naturally focus on the lives and wellbeing immediately at stake. But what happens during a crisis can also affect what happens when the situation returns to normal, or a new normal. So Tornado politics influence longer term systemic politics around which there is no single decision or pre-defined sequence of decisions. The British response on Fukushima may have informed Japanese willingness subsequently to invest in British nuclear power new build at Hinkley Point, which in turn was an important moment in the evolution of policy on nuclear energy in the UK. Meanwhile, the emergency was also linked to the change in German energy policy that led to the decision to phase out nuclear power.

Decision-makers and their advisers often face the choice between a reasonably certain answer to a partial question and a very uncertain answer to a "better" and more comprehensive question. Ultimately the natural and physical systems considered to surround any policy decision could be extended indefinitely outwards to encompass just about everything on the planet (or extended inwards to quarks, probably). It is also inevitable that sometimes the policy "system" is, at key moments such as in Cabinet government or during budget negotiations, so broad that the decision on one area of policy is made as part of a set of negotiations or trade-offs against a completely unrelated set of policies. It can be hard for those involved in developing specific policy option, whether scientists, officials, citizens or even sometimes Ministers, to see these trade-offs happen. Being resilient to such moments is often a requirement for staying at the interface of science and policy.

Staying with the more traditional physical and temporal scales of policy debate, it may be helpful for scientists to make more explicit why and how they are defining their systems. To caricature: for many scientific disciplines, abstracting elements of a natural system and building models of it in order to enhance understanding is so much at the heart of how they work that it can feel like a moral imperative. The system may be defined by what can be modelled or considered traditionally within a discipline or a group of disciplines used to working together.

In the physical sciences, therefore, there is little reason for the system being studied to coincide with the with policy question. Arguably, one of the challenges of climate science and policy is the fact the evolution of atmospheric and other earth sciences, based on the forms of observational evidence and physical theories available, led to models for which a key indicator of outcomes was the measure of global average surface temperatures. This measure is not well designed to inform public debate and public decision-making, at least at the regional, national or local levels which are where most policy levers lie. In its extreme form the mismatch between what needs to be modelled for a particular purpose and what can most easily be modelled scientifically takes the form of the old joke amongst physicists that they can model anything, provided it can be assumed to be spherical and in a vacuum and so taking the simplest form possible for the purposes of applying the laws of physics. Economists' assumptions of equilibrium, perfect information and human rationality are a less extreme form of the same fundamental approach to the challenges of understanding, explanation and prediction.

These questions of how far evidence that informs policy is framed by the observations and theory available, but developed for purposes other than policy, applies not only to independent research but also to the considerable investment in scientific evidence by industry and by civil groups. It may be the case that, like looking for your keys at night on the patch of ground illuminated under a lamp-post, the availability of the evidence frames the options that may be considered. One of the many joys of science, however, is that it may well be the case that while public debate focusses on the area under lamp-post A, some scientists exploring the area under lamp-post B apparently positioned for reasons to do with basic scientific curiosity, or placed to illuminate a completely unrelated set of questions, come up with a wholly new set of ideas for tackling the problem at hand. The question of how to ensure that the potential solution-holders and potential question-owners find each other is explored further in Chap. 5.

The issue of placing the lamp-posts becomes increasingly important as the power of data science increases the range of policy questions to which quantitative modelling may be applied. A particular challenge is that the data may have been collected for reasons totally unrelated to the policy issue and therefore be biased (in the sense of skewed, rather than prejudiced) with respect to the policy concerns. Or it may be that increasing amounts of publicly valuable data are wholly owned by the private sector.

For policy-makers, the system that defines their decision-making may be set by several factors. It may be the history and the pathway of the debates, the pre-existing pattern of stakeholder interests, or the sequence of recent events and social amplification of selected aspects (Pidgeon, 2003). Sometimes the framing is determined by a much wider political narrative: for example, concerns about AI and automation being framed by impacts on jobs that are also part of a wider set of concerns about globalisation and inequality.

Policy-makers' systems may be set by their departmental boundary, or their access to the levers of influence. Scientists often have the luxury of being able to define the system in a way they believe may be most tractable to science, while policy-makers have to fight very hard to frame the issues in a way that makes them tractable to policy, and sometimes simply cannot achieve it.

The policy framing may also be influenced by factors such as the tendency for narratives to focus on individual or concrete problems at the expense of the larger system. So, for example, the long-running disputes about whether to cull badgers to help manage the risks from bovine TB exemplify the reality of policy debates' tendency to be framed around, in this instance, charismatic entities and relatively simple narratives and choices. The longer term solutions require, amongst other things, consideration of the much less tangible issues of biosecurity in farms, agricultural business models and the management of populations of a wide range of wild and domestic animals.

These issues of framing the moment of decision: the debate, the systems and the nature of potential solutions, are also helpfully explored in the many discussions of public and private sector "wicked problems" (see, for example Hulme (2009)). Wicked problems have many causes, are hard to describe, don't have a single answer that will satisfy all who have a stake in them and, typically, each step to tackle one aspect creates its own new challenges. Wicked problems have a counterpart in the concept of Post-Normal Science in which the uses of different forms of evidence are themselves contested due to different norms and values (Funtowicz, 1993).

At this point however, it may help to return to the craft skill of simply getting things done. The scientist who, on their first encounter with a Minister, begins only with "well Minister, it's very complicated and very uncertain" or the scholar who seeks solely to further "problematize" the issue, is unlikely to have a second encounter. Politics can, or rather in many cases has to, get on without science. In this moment of practice, a skilful adviser can acknowledge complexity and uncertainty while working hard to show that their aim is to enable the Minister to make a well-informed, reasoned and defensible decision.

MODELS OF THE POLICY-MAKER AND THEIR ENVIRONMENT

Whether the situation reflects Tornado, Abortion or Ecstasy politics, theories of policy-making, such as those outlined by Paul Cairney (2016) and others, outline basic models of the policy-making environment and of the individual within it. Cairney speaks of the building blocks of the environment as being the wide range of actors, the many levels of government, differing rules and norms, the effects of networks between policy-makers and other actors, a tendency for certain framings to dominate discussion, and the reality of changing policy conditions and of events that shift the policy-maker's attention and priorities often and short notice and unpredictably.

These realities, and the limits to human cognition and scientific evidence, mean the individual policy-maker often has to frame, evaluate and react to problems or opportunities with only partial information or evidence of any kind. They must act without being able fully to understand what the problem or opportunity actually is, and certainly without the level of insight into potential outcomes of their decisions that would be required in many other professional settings such as business, where the decision-maker usually has somewhat more control over their environment.

To survive and to get things done in this situation essentially requires the decision-maker to operate in two modes at once. The "rational", as described by Cairney, includes pursuing clear goals and prioritising certain kinds of information. The "irrational", draws on emotions, gut feelings, deeply held beliefs, habits and the familiar. This dualism has similarities to Daniel Kahneman's descriptions of the fast intuitive and slower deliberative modes of human thought (Kahneman, 2012). But there is a difference, because although the policy-maker may make decisions like any other human being, they will have to defend them subsequently in ways that most of us do not. In defending them, they will have to deploy arguments, or narratives, that seek to persuade or defend the "irrational" aspects to their decisions. In some ways, that means applying slow modes of thought to justify fast decisions.

The distinction between rationality and irrationality comes loaded with preconceptions that differ by audience, and all the terms are loaded. To avoid giving apparent primacy to rationality by referring to its complement as irrationality, this book will refer to the complementary quantity as emotion or sentiment.[1] One thing should be obvious to anybody who has ever met one, and that is that scientists are just as emotional as any other human being; but the entire scientific system is designed to channel the passionate energies of researchers into paths of creation and testing that

---

[1] For the purposes of this book, sentiment and emotion are broadly used as synonyms, as are cognition and rationality.

produce knowledge that transcends the individuals who first expressed it. Similarly all citizens are scientists, in the sense that at least for some of the time they cannot help but be curious, imaginatively form hypotheses, observe the world, and learn from that experience.

When it comes to the craft skills, the distinction between sentiment and rationality is perhaps less helpful. Because policymakers need to be able to operate in both modes, a scientist could regard it as perfectly acceptable, even as a duty, to produce rational evidence and then to illustrate it, or to work with others to illustrate it, with sentimental narratives. These might be case studies or personalised accounts, deliberately designed to be draw attention to what they consider to be key aspects of the evidence. The danger to well-founded public reasoning emerges when the sentimental narrative is taken to outweigh the evidence, or distract from it. Chapter 3 discusses the supporting and distorting roles of narratives in more depth.

ROLES OF THE SCIENTIST

Pielke also introduced a helpful framework for all parties to consider what role a scientist is adopting in any given moment. In The Honest Broker (Pielke, 2007) he outlines four roles. The Pure Scientist focusses on research without regard to its policy implications. The Science Arbiter seeks to provide authoritative scientific evidence, usually when requested by policy-makers and only well within the relevant scientific disciplinary field. Many scientists on technical scientific advisory committees adopt this role. The vast number of decisions inside government are simply about getting business done and are not controversial, so experts of one sort or another, including scientists and practitioners, are frequently in Arbiter mode.

The Issue Advocate goes further into the policy process, focussing on what the research means for a particular agenda. Over much of the second half of the twentieth century many climate scientists were Issue Advocates, reflecting their views on the scale and urgency of the action required in light of the emerging evidence.

The "Honest Broker of Policy Alternatives" engages with the policy process, seeking to clarify or sometimes to expand the range of options available to the decision-maker. As a concept the Honest Broker helped crystallise a shift away from issue advocacy in some contexts and towards a greater recognition that scientific evidence is often only one aspect of the elements that should inform public decision-making. An optimistic interpretation is that the shift was helped, in the UK, by wider recognition that public reasoning should indeed be informed by science, and by the emergence of more scientists willing and able to help make this happen.

The same scientist can adopt different roles in different contexts, depending on the nature of the science and of the policy need. Being self-aware, deliberate and transparent about what roles are being adopted is likely not only be good for policy outcomes, but also better for the long term conduct and perceptions of science.

Looking more granularly, the Chief Scientific Adviser of New Zealand, Sir Peter Gluckman, articulated 10 types of activity that may be included within a scientific advisory role (Gluckman, 2014). These included evaluating pre-determined policy options, identifying and managing risks, and creating or illustrating new possibilities. While several governments and cities have Chief Scientific Advisers or equivalents there are, of course, many actual and potential arrangements for the flows of evidence and for conversations between decision-makers and scientists. Such arrangements vary with the wider political and scientific context, and the needs and characteristics of the time (see Wilsdon and Doubleday (2015) for a discussion of science advice in the European Union, for example). Even within the UK's regime, different CSAs have chosen to weight different parts of their role and it will no doubt continue to evolve. For the purposes of this book, formal advisory roles are particularly interesting primarily because they exemplify aspects of the challenges and opportunities facing any individual scientist or intermediary seeking to provide advice or to enable others to do so.

BRINGING SCIENCE AND SCIENTISTS TOGETHER TO SERVE POLICY

To any policy question of significance, more than one academic discipline applies. This means that the challenge of brokerage doesn't just apply to the role of the individual scientist with respect to their core discipline at the policy interface. It is also essential to the task of bringing multiple disciplines to bear. This form of inter-disciplinary brokerage is itself an expertise. It requires judgement about what disciplines to bring in, how and through whom. These issues are discussed further in Chaps. 4 and 5 and, like so many other aspects of science in government, insightful approaches require the application of both conceptual frameworks and craft skills.

To be very clear, the starting point must be for all forms of science and scholarship to be in the frame. This was strikingly illustrated by the case of advice in a second civil emergency, that of dealing with the Ebola epidemic of 2014. Here, as for the Fukushima emergency, a SAGE was established to advise COBR. However it rapidly became obvious that, to be effective, efforts to tackle the outbreak had to be informed by expertise of anthropologists and social scientists. These researchers helped ensure that, for example, the full significance of burial and other important indigenous practices were properly recognised and integrated into the operations (House of Commons Science and Technology Committee, 2016).

The practice of drawing together insights from different disciplines is often referred to as evidence synthesis. Synthesis can happen in a matter of seconds in an expert's mind, in a matter of hours and days during a crisis as with Ebola or Fukushima, or over months or years as with various established systems such as the Cochrane medical reviews or the Intergovernmental Panel for Climate Change (IPCC), which oversees a comprehensive synthesis process with suites of reports every five or so years.

High quality synthesis requires careful selection of the disciplines and sources of expertise, rigorous quality assurance and challenge mechanisms, and intellectual independence, coupled with close engagement with the intended audience (Donnelly et al., 2018). At its best, it helps the policy-maker and publics navigate different types of evidence and make better sense of stakeholder arguments based on cherry-picked evidence (for a good account of the many forms of abuse of evidence see Mark Henderson's The Geek Manifesto (2012)). For example, the Oxford Martin School Restatements tackled subjects controversial in UK politics such as bovine TB (Godfray et al., 2013) and the effect of neonicotinoids on pollinating insects. By setting out and assessing different types of evidence they helped illuminate the different types of knowledge, and of uncertainty associated with each, relevant to the policy question. The Royal Society's judicial primers represent another example of evidence synthesis, targeted in this case at the courts and created in collaboration with bodies such as the Judicial College (The Royal Society and others, 2017).

Public reasoning would often benefit from the provision of carefully synthesised and easily accessible evidence but there are currently relatively few incentives to provide it. Academics are primarily rewarded for creating new knowledge within their disciplines and, in the absence of reliable access to synthesised evidence in most policy domains, policy-makers do not learn to ask for it.

However there are several factors that mean the situation may improve. Despite the perception that policy-makers' needs of researchers move faster than the science, which is often true, it is also true that many deep rooted policy issues stay the same for very long periods of time. This means the research questions in which policy-makers in those areas are likely to be interested also remain the same, and gives time and space to establish individual research programmes and to build mechanisms for synthesis. Increasingly, too, researchers within a field need more systematically to access the knowledge available within it or within adjacent fields, in order to inform their future research. So elements of synthesis can be valuable for the research community too.

END NOTE: CRAFT AND CURATION

There have been several references to craft skill in this chapter. That reflects the theme throughout this book that the practice of enabling scientific knowledge to take its place within government is ultimately a skill that combines analytical thinking and conceptual insight with judgement, creativity, instinct, empathy and attention to practical detail. To use a different metaphor, getting the Minister and the Nobel Laureate into the same room for a productive conversation is an act of curation, as well as craft, and this is discussed further in Chap. 5.

REFERENCES

Cairney, P. (2016). *The politics of evidence-based policy making*. Palgrave Macmillan.
Donnelly, C. A., Boyd, I., Campbell, P., Craig, C., Vallance, P., Walport, M., et al. (2018). Four principles to make evidence synthesis more useful for policy. *Nature, 558*(7710), 361–364.
Funtowicz, S. O. (1993). Science for the post-normal age. *Futures, 31*(7), 735–755.
Gluckman, P. (2014). Policy: The art of science advice to government. *Nature, 507*, 163–165.

Godfray, H., Donnelly, C., Kao, R. R., Macdonald, D., McDonald, R., Petrokofsky, G., ... McLean, A. (2013). A restatement of the natural science evidence base relevant to the control of bovine tuberculosis in Great Britain. *Proceedings of the Royal Society B, 280*, 20131684.

Henderson, M. (2012). *The geek manifesto*. Bantam Press.

House of Commons Science and Technology Committee. (2016). *Science in emergencies: UK lessons from Ebola*. London: The Stationery Office Limited. Retrieved from Publications, Parliament.

Hulme, M. (2009). *Why we disagree about climate change: Understanding controversy, inaction and opportunity*. Cambridge University Press.

International Atomic Energy Agency. (2015). *Director General's report on Fukushima Daiichi accident*. International Atomic Energy Authority.

Kahneman, D. (2012). *Think, fast and slow*. Penguin.

May, T. (2013, July 9). Letter to ACMD on control of khat. *Gov.uk*. Retrieved April 13, 2018, from https://www.gov.uk/publications/letter-to-acmd-on-control of khat

Nutt, D. J. (2009, January 21). Equasy – An overlooked addiction with implications for the current debate on drug harms. *Journal of Pscyhopharmacology, 23*(1), 3–5.

Pidgeon, N. (2003). *The social amplification of risk* (N. Pidgeon, R. E. Kasperson, & P. Slovic, Eds.). Cambridge University Press.

Pielke, R. A., Jr. (2007). *The honest broker*. Cambridge University Press.

The Royal Society and the Royal Society of Edinburgh in conjunction with the Judicial College, the Judicial Institute, and the Judicial Studies Board for Northern Ireland. (2017). *Forensic DNA analysis*. The Royal Society.

Thomas, G. A. (2011). Health risks from nuclear accidents. *Annals of Academy of Medicine, 40*(4), 40.

UK Government. (2010, March 24). Principles of scientific advice to government. *Gov.uk*. Retrieved April 9, 2018, from https://www.gov.uk

UK Government. (2012, April 5). Government office for science, civil contingencies. *The National Archives*. Retrieved April 9, 2018, from http://webarchive. nationalarchives.gov.uk

Wilsdon, J., & Doubleday, R. (2015). *Future directions for scientific advice in Europe*. Centre for Science and Policy.

CHAPTER 2

How to Express Risk, Confidence and (Un)Certainty

RISK MANAGEMENT IS CORE TO GOVERNMENT

Manifestos and Ministerial speeches typically promote aspirational policies and new ideas but a very great deal of the daily work of government is concerned with managing risks. Risk management here refers to the practice of dealing with reasonably well-defined risks which materialise rapidly. Risk itself is a measure of the harm caused by the potential event multiplied by the likelihood of that harm occurring (see, for example, Lofstedt (2011). This section also draws extensively on Craig (2018)).

The mechanism for the advice given during the Fukushima emergency is an example of the application of the UK's systematic approach to the management of major risks. Every government department, like every major business and University, has a risk register that includes financial, operational and reputational risks. But the most visible and consistent expression of risk at national level is the National Risk Register (NRR). The Register is the result of systematic and regular exercises led by the Cabinet Office with the aim of improving national risk management and it includes risks such as pandemics and major infrastructure failure. It provides a public resource to enable emergency services, local authorities and others to be better prepared for emergencies (UK Government, 2017). Within the NRR the principal risks are summarised in a matrix of probability against impact. The identification and examination of many of the risks is informed by operational and scientific expertise, with the latter largely coordinated by the Government Chief Scientific Adviser (GCSA).

© The Author(s) 2019
C. Craig, *How Does Government Listen to Scientists?*,
https://doi.org/10.1007/978-3-319-96086-9_2

15

Not all these risks have crystallised in living memory. For example, the Register includes the possibility of a major solar flare, which would disrupt electrical systems, and there has not been a major solar flare since 1859. But in all cases the assessors either have access to historic records of the events or, in the case of risks such as those from major flooding or pandemics, experience of smaller versions of the same type of hazard.

Once risks get on to the Register, there is a system for ensuring responsibility for planning and acting to mitigate the risk. In operational terms, the task of achieving preparedness is made slightly simpler by the fact that many of the potential responses, from dealing with rubble to the loss of communications, are common to risks that themselves have very different individual causes.

So the purpose of defining and prioritising the risk is to inform the actions. Meanwhile the government's or agency's marginal spend or marginal moment of organisational focus on the risk competes with their spend or focus on the much more obvious needs of the policy and operations of the day. This makes it both important and challenging to create and present evidence about events that have never happened and may not happen and which can be hard to characterise let alone to quantify, in ways that best inform judgements about proportionate investment of resource.

In some ways, this is the body politic mimicking the brain. Seeking knowledge carries psychological danger (Eiser, 2004). Human brains are not designed primarily to find out the truth about things but to keep us alive and, although these purposes overlap, they do not always align perfectly. So in policy, and implicitly in public discourse, the question of why we are discussing a risk matters greatly. The question the decision-maker, whether a policy professional or a member of the public, is likely to be asking very early on is "what does this mean for me, and what can I do about it".

Through the NRR and other mechanisms the UK's formal risk management processes are well established compared to those of some national governments (Hemingway & Gunawan, 2018), but there is further to go and they continue to be developed. For example, the NRR does not consider the cascade, convergence or correlation of risks. Fukushima was an example of these, as the earthquake, tsunami and nuclear emergency could have been considered as independent risks. The impacts of the UK's extreme weather events of 2014 combined the consequences of storms and flood water in the Somerset Levels and elsewhere, with disruption to road and rail transport, power supplies and the potential for infectious diseases in animals.

The NRR also cannot deal fully with High Impact Low Probability risks which include natural disasters such as unprecedented volcanic eruptions and hard to imagine risks in the category sometimes referred to as Black Swans (Taleb, 2007). Therefore, under Sir John Beddington, the Government Office for Science reviewed the government's approach to High Impact Low Probability risk management (Government Office for Science, 2012). That report highlighted the way in which considering such risks requires insights from a very wide range of disciplines. For those managing them, it becomes important to adopt multiple approaches, such as identifying early warning signals, including "near misses" (as happens with aircraft and some financial safety regimes) and explicitly considering linking or compounding risks.

The report also noted that the task of focussing on moments of sudden and extreme risk, while very important, needs to be placed in the wider context of systemic and long term issues. For example, measures of the risk to a coastal area from a future major storm surge depend not only on the nature of extreme events but also on long term trends in sea level, coastal ecology and human activity.

People and organisations tend to ignore problems that are too complex to fit easily into current frameworks of accountability and behaviour. In the most extreme form they exhibit wilful blindness, a popular term to describe the situation in which someone could have known and should have known something but, either consciously or unconsciously, did not find it out. Wilful blindness is a contributor to corporate failures such as that of the energy business Enron and to accidents such as that at the BP Deepwater Horizon oil rig. Some of its causes relate to the very great systemic incentives on individuals to keep to the familiar, to retain a positive self-image and the social networks that reinforce that, and so to avoid evidence that will disrupt any of these (Heffernan, 2011).

The individual scientist taking part in discussions about extreme risk has to come to terms with several problems. They are likely, by definition, to have to give at least temporary credence to notions which challenge current thinking and to possibilities well beyond the scope of historic observation. Framing the risk is a multi-disciplinary challenge and multi-disciplinary conversations invariably require participants to do things that are often hard for scientists to do, such as simplifying their language, or revealing their ignorance of other specialist areas.

Overcoming these problems is in part a matter of curation. It requires careful techniques for making space and time for groups to engage in creative thinking and testing assumptions. These can be drawn from practices in innovative businesses or parts of government, defence settings, and theories of decision-making and creativity. When working with scientists and scholars, however, it is particularly important to remember that the situation may need some deep domain knowledge buried in the head of the expert in a single domain who is personally the least comfortable speculating or debating.

RISK AND INNOVATION

Risks are popularly considered to be negative but many professional fields speak of "upside", as well as downside, risks. Risk and innovation are intertwined in this way. All innovation carries risk but, at least in the sense of change and growth, innovation is essential. So both acting and not acting carry risks.

Science appears on both sides of the innovation equation. Scientific knowledge can create innovation and can also be used to inform responses to it. Innovation comes from new ideas and new possibilities, enacted through changes to systems and behaviours. Social innovation would continue if the physical sciences stopped today. But much of the discussion of innovation in government relates to new technologies that are themselves at least partly the product of science. The policy challenge is to enable new technologies to disrupt existing social and physical systems in ways that enable as many of the benefits of innovation to be delivered as possible, while managing the downside risks.

In these contexts considering risk rigorously makes the decisions more robust (Walport & Craig, 2014). Such framings need to consider action and inaction, costs and benefits and, ideally, multiple pathways to the desired outcome. The public disputes around the case for vaccination for diseases such as measles in the UK highlight the need to consider the consequences of choosing not to act. Visualising the consequences of inaction in this setting requires the creation of a compelling public narrative for the case for vaccinating against a disease that few have experienced. Of course, evidence is always contingent and uncertain, and so are people's responses, an issue which is explored in more depth in Chap. 4.

In the case of GM crops, European public debate was regarded by many to have focussed nearly exclusively at times on the extent to which GM crops might harm the environment or human health, when the reality was that there were also significant issues at stake concerning the wider systems governing the technologies and their uses. Questions about the science were embedded in concerns about the roles of global businesses in developing and implementing the relevant technologies and business models, or with the social and cultural meanings of the countryside, or of food production and consumption.

Unreflective framing can lead to confused or misdirected debate and effort. Then the scientific and technical questions of risk may act as a lightning rod attracting discussion away from other, perhaps more difficult and sometimes ultimately more significant, questions. It can leave scientists and science advisers pouring potentially disproportionate effort in trying to refine the evidence on a point of harm that was not, and never would be, capable of resolution by the evidence being considered and which was not what many people cared most about anyhow.

Regulators of new technologies often operate in an asymmetry in which the benefits of a future good are likely to accrue to people who do not yet know that they risk missing them. At the same time, the disbenefits are often visible and will affect existing identifiable stakeholder groups. So in some cases a regulator may feel they are more likely to be penalised for letting a bad thing happen than for neglecting to enable a new form of good. All of these considerations need to be balanced against the reasonable grounds for precaution in novel technologies and, ultimately, the precise mechanisms for acquiring evidence and making and implementing judgements often need to be highly context specific.

THE POWER OF NUMBERS

There are many insights into the major question of how to express uncertainty (Blastland & Spigelhalter, 2013). The perception, often the reality, being that scientists are comfortable with it and deploy a variety of forms for expressing it, while politicians tend to want certainty and feel the media demands it. There was a small moment of celebration in the analytics team of one government department, for example, when a Minister first agreed to provide an answer to a Parliamentary Question in the form that included the confidence intervals around the requested and politically charged number.

Often the situation starts with just such a politically contested number. Numbers are powerful; totemic. A good number expresses a deep truth about a complex system like, say, the climate or patterns of crime. It can move around quickly, and spread understanding amongst decision-makers and the public. As a target, a number can motivate and galvanise action. The corollary is that a poorly founded number is just as likely to be attractive and long-lived, so it is of course essential to get them right.

The statement that the world needs to hit a target of reducing carbon emissions by at least 80% from 1990 levels by 2050 in order to avoid excessive damage from climate change, is (relatively) straightforward. So is the target of keeping global average surface temperature increase to less than 2 degrees Centigrade. Here, the numbers are powerful signalling devices and are the product of the processes both of science and of policy. It does not matter for the most part, given the size of the physical and socio-political uncertainties, whether 2 should be 2.01 or 1.99. In these circumstances a very precise numerical statement will inevitably be so qualified and bounded that it requires domain expertise to interpret the caveats, or it will be reduced to applying to only part of the system that matters in policy terms. In either case, its power outside of the expert group is lost.

Numbers can also rapidly give heuristics for thinking about complex systems. In an hypothetical illustration: to be told that within a specified system 30% of men have a criminal conviction by age 30, but 80% of crime is committed by 20% of offenders, conveys some intuitive sense of the landscape of crime, that might helpfully direct the next stages of exploration.

It used to be the case that scientists picked their numbers without much thought to the emotional or narrative impact. But the increasing range of evidence from behavioural and social sciences, including experimental psychology, means the choice of expression is itself something that can be approached more expertly.

The challenges of communicating risk through numbers, particularly in the contexts of climate and weather and of human health, are being studied extensively (to pick one example from many, see David Spiegelhalter's work on the health impacts of air pollution (Spiegelhalter, 2017)). So it is possible to understand more about what people think about numerical information about risk in a particular context and hence to understand more about how such information might affect their opinions or actions. For example, for certain types of risk, people are typically more confident in the information if the risk is given as a range of values rather than a point prediction. In some contexts their interpretation of the likelihood of an event taking place is influenced by the severity of the event: depending on the circumstances they tend to think snow flurries are more or less likely than a hurricane, even when given the same numerical probability (Harris & Corner, 2011). Similarly, it is possible to assess what numerical probabilities people associate with a word such as "likely" or "very likely", in a particular context.

We don't yet have good models to help create fully generalizable models of the interpretations of risk or particular numerical descriptions of uncertainty across different areas of science in public debate. But we do know that these are matters on which it is possible to use more than intuition.

THE INFLUENCE OF IMAGES

While policy discussions return repeatedly to numbers and words, images matter too. Much of science creates and uses images (Secord, 2014). They might be detailed observational drawings of fossils, creatures or rocks. They might be seismographs needing interpretation, or images created from data or models such as X-ray photographs of galaxies rendered in colours visible to the human eye, maps of surface temperature over time, of the risk of flooding, or displays of temporal or spatial patterns in types of crime.

Much public debate is mediated through still or moving images; of tsunamis or volcanoes, of drought, of a polar bear on an ice floe, a turtle caught in plastic waste, or of a malaria victim. The images that draw attention may also be representations of data, such as the "hockey stick graph" of historic global temperatures or the controversial "bell curve" distribution of IQ measurements in a population. The Foresight future flood risk maps of England and Wales discussed in Chap. 3 helped draw attention to the work and helped decision-makers find a way in to engaging with it more fully.

There are also cases under which the choice of image directly affects individual and collective decisions. First, and where intended to convey complex information to inform decision-making, they may operate differently from numbers even if derived in the same way. So inhabitants of towns in Texas were more likely to evacuate from an area at risk of being in the path of an extreme weather event if they saw the projected height of the floodwater in a graphic in which it was shown lapping against the their buildings, than if they were given the same information in numbers. Virtual reality displays have helped public preparations for bush fires in Australia.

Just as the introductions of writing and of mathematical tables increased humans' memory and computational power, so visualisations may extend our analytical and decision-making power. But increasingly the advent of massive amounts of data about complex systems means that the task of creating a skilful visualisation can be as significant as the challenge of doing the science in the first place (McInerny et al., 2014).

Insights into how to convey complex information to inform decisions are growing. Some may come from the computer gaming industry, or the military where the task of enabling a soldier or pilot to operate all the available equipment is increasingly constrained by the need to limit and channel the volume of information that equipment creates. Many new insights about corporate and public sector strategy originated with strategic approaches to business or war, and both may drive further insights on the best ways to communicate complex information.

For now the craft skill of providing graphic information to decision-makers includes extreme attention to detail, and improvements may be relatively low tech. It is still possible to find the most sophisticated sources of information being provided through interfaces which, for example, have different scales and uses of colour, making it more difficult for the user to assimilate them accurately.

THE IMPORTANCE OF WORDS

There is, of course, a powerful third form of communication alongside numbers and images, which is words. Words gain a special status in public decision-making partly because of their relationship to Acts of Parliament, regulation and the way law is expressed and enforced. (As an aside, the structure of logic and flow in the words expressed in a Bill going before Parliament can show a striking similarity to the structure of code in early computer languages such as Fortran, presumably demonstrating an underlying equivalence in the task of using words for setting specific goals for the operation of complex systems.)

Some would argue that, for the purpose of well-founded public reasoning, there are circumstances where only words will do (Jasanoff, 2012). Depending on the precise context, this is often true, but words also typically fail to convey the intuitive understanding of complex systems which requires something more like the engagement with models or multiple narratives which is discussed in Chap. 3. The extended and sustained efforts to engage with the implications of climate change illustrate this line of argument: for multiple individuals and publics each in their own way to grasp the essence of what is sometimes reduced to single numbers, words or images, is in fact the task of activities from computational modelling and thinking through objects, to film, art and dance.

Back with the craft skill lies the challenge of deciding when and how to define significant words. It is helpful that the only language which different academic disciplines can share is also that of the Minister or the public, so translation only has to be done once. When the scientific disciplinary terms are incomprehensible to those outside the discipline, the need for translation will be obvious. The biggest misunderstandings between disciplines or between scientists, scholars and publics are more likely to come about because the words appear familiar and are therefore used without the users realising they mean different things.

So, for example, "bias" in data to the scientist means a data set which does not accurately represent the target population: for some purposes, it is simply a challenge to be solved. To the generalist it typically means "prejudice", in that someone with agency in the situation has or will act preferentially towards one group or one objective over another. "Theory" has public overtones of "theoretical", meaning the opposite of practical, and with a high degree of uncertainty. To a scientist it may refer to a set of rules that have been in widespread use for centuries, making them about as certain as anything can be (Somerville & Hassol, 2011).

These misunderstandings can take time to recognise. In one event involving computer scientists and neuroscientists who had not previously engaged with each other, the discussions of memory were many hours advanced before it became apparent to all that they used the word very differently. To the former group memory was essentially a quantity that they wished to grow under all circumstances, never to lose and always to be able to access rapidly and without variation (the right to be forgotten had to be written in to law). To the latter, it was a finite and fluid quantity where loss, in the sense of forgetting, was an integral part of learning and improving.

Behavioural sciences are beginning to increase knowledge about how words, the process of naming things, creating phrases or full narratives, affect individual decisions. Meanwhile, as discussed further in Chap. 3, the times seem ripe for digital humanities and social sciences to make further breakthroughs into understanding how such practices relate to the broad currents of public debate.

Science is not immune to the mesmerising effect of a sticky word or name. Terms invented to describe aspects of science or technology for which public debate had previously not needed words at all can grow out of technical spheres and be more widely appropriated. The use of spam as a new term to describe unwanted email traffic grew rapidly. This was probably both because of how it sounds and because of its associations with Monty Python sketches and the original tinned meat which meant it had tribal and emotional heft (Keats, 2010).

In the urgency of a public policy setting, it is important both to be alert to damaging misinterpretations, and not to get caught up in unwinnable definitional arguments between groups. In particular, sometimes it is necessary to assert strongly that the definition of a word needs to be for the purposes and with the intended audience at hand. It is not always an intrinsic quantity waiting to be uncovered by rational argument nor a quantity that can be fully imposed by one group on another.

Meanwhile, back in the room with the Minister and the Nobel Laureate, each will have an individual preference for the ways of absorbing information, just as all humans have preferred ways of learning. The wise broker either knows which ones to use, or uses a little of all.

ERROR, UNCERTAINTY AND DECISION-MAKING

In many cases there is a powerful asymmetry in politicians' and scientists' orientation towards errors. Politicians are likely to be concerned with avoiding the "Type I" error of wrongful inaction. Politically and publicly choosing not to act in the face of perceived need or opportunity, whatever the weakness of the evidence, can be hard. However scientists may be more concerned with the "Type II" error of wrongful action or reporting a finding where the finding is not robust. For scientists, avoiding Type II errors is a form of caution burned in by a system that requires a scientist to be very confident indeed before they publish a result in front of their peers.

The use of numbers is valuable for mediating between forms of evidence but can mask very different causes and types of uncertainty. So, for example, uncertainty about the extent and nature of climate change impacts arises from many different sources. Uncertainty in the outcomes of models of physical systems may derive from the quality of historic data, from the assumptions underpinning the model, or from computational limitations, amongst many other things. Uncertainty in the wider evidence may derive from models' relationships to the relevant real-world system, including the deep uncertainties in social, economic and demographic futures. Only some of these sources of uncertainty are capable of being remedied in the near or medium term due to new observations, technologies or insights. Some forms of uncertainty may be intrinsic to the system, due to its complexity. There may simply be things that are presently unknowable: they are mysteries, rather than puzzles that can be solved.

Rational decision-making theory would imply that the decision-maker would not act under any significant degree of uncertainty (Oreskes & Conway, 2010). However, in policy-making, once the problem or opportunity has been identified the incentives and pressures are more usually in favour of avoiding the Type I error, by taking a decision to act. In the White Queen moment, the pleasure of deciding to act is felt now while the actions themselves, with their real rather than their anticipated effects, take place in the future (Caroll, 1871).

REFERENCES

Blastland, M., & Spigelhalter, D. (2013). *The norm chronicles: Stories and numbers about danger and death*. Profile Books.
Caroll, L. (1871). *Through the looking-glass, and what Alice found there*. Macmillan.
Craig, C. H. (2018, May 5). Risk management in a policy environment: The particular challenges associated with extreme risks. *Futures*.
Eiser, J. R. (2004). *Public perception of risk*. UK Office of Science and Technology. Retrieved April 13, 2018, from https://pdfs.semanticscholar.org
Government Office for Science. (2012). *Blackett review of high impact low probability risks*. UK Government Office for Science.
Harris, A. J., & Corner, A. (2011). Communicating environmental risks: Clarifying the severity effect in interpretations of verbal probability expressions. *Journal of Experimental Psychology, Learning, Memory and Cognition, 37*(6), 1571–1578.
Heffernan, M. (2011). *Wilful blindness*. Simon & Schuster.
Hemingway, R., & Gunawan, B. (2018). The natural hazards partnership: A public-sector collaboration across the UK for natural hazard disaster risk reduction. *International Journal of Disaster Risk Reduction, 27*, 499–511.
Jasanoff, S. (2012). *Science and public reason*. Earthscan. Routledge.
Keats, J. (2010). *Virtual words: Langage from the edge of science and technology*. Oxford University Press.
Lofstedt, R. (2011). Risk versus hazard – How to regulate in the 21st century. *European Journal of Risk Regulation, 2*(2), 149–168.

McInerny, G. J., Chen, M., Freeman, R., Gavaghan, D., Meyer, M., Rowland, F., ... Hortal, J. (2014). Information visualisation for science and policy: Engaging users and avoiding bias. *Science, 29*(3), 148–157.

Oreskes, N., & Conway, E. M. (2010). *Merchants of doubt.* Bloomsbury Press.

Secord, J. A. (2014). *Visions of science, books and readers at the dawn of the Victorian age.* Oxford University Press.

Somerville, R. C., & Hassol, S. J. (2011, October). Communicating the science of climate change. *Physics Today.* Retrieved April 22, 2018, from https://www.climatecommunication.org

Spiegelhalter, D. (2017, February 20). Does air pollution kill 40,000 people each year in the UK? *Winton Centre.* Retrieved May 7, 2018, from https://medium.com/wintoncentre

Taleb, N. (2007). *The black swan: The impact of the highly improbable.* Random House.

UK Government. (2017). The national risk register of civil emergencies. *Gov.uk.* Retrieved from https://www.gov.uk/government/publications

Walport, M., & Craig, C. (2014). *Annual report of the Government Chief Scientific Adviser 2014. Innovation and risk: Managing risk, not avoiding it.* Government Office for Science. Retrieved April 13, 2018, from https://assets.publishing.service.gov.uk

CHAPTER 3

How to Think About the Future

MODELS LINK PAST, PRESENT AND FUTURE

Observational evidence is always about the past. Policy-makers want to know the consequences of their decisions for the future. Some of the biggest misunderstandings between scientists, scholars and policy-makers in both private and in public debate come about because of confusion about the bases of the relationship between the two.

One link is through models. "Humans are natural modellers—we carry models of our world in our minds. Our memories are significantly comprised of a mental model of the world in which we live, and our personal history of our experiences within that world. We navigate by means of maps: mental maps and the physical maps that we create" (Government Office for Science, 2018).

These might be intuitive and personal mental models about how some aspects of the world work, or they might be explicit and shared constructs. The constructs might in turn be expressed in a variety of forms. They might be architectural models such as the physical model of St Paul's Cathedral constructed for Sir Christopher Wren, or today's 3-D computer designed and sometimes 3-D printed models. They include wonderful examples such as the physical hydraulic model of the macro-economic relationships between stocks and flows in an open economy created by William Philips in 1949. They may be conceptual models or they may be intuitive models of how families and social groups work, underpinning narratives such as, say, Persuasion by Jane Austen.

© The Author(s) 2019
C. Craig, *How Does Government Listen to Scientists?*,
https://doi.org/10.1007/978-3-319-96086-9_3

In broad terms, models are used for at least five purposes: prediction or forecasting, the explanation or exploration of future scenarios, understanding theory, illustrating or visualising a system, and analogy (Calder et al., 2018). They are always in some sense fictions, but they are fictions that simplify and abstract important properties of the actual object or system being modelled, to create insights or outputs useful for the purpose at hand (see for example (Currie, 2017)).

In science, computational models form the basis of understanding of the largest systems, such as galaxies, and the smallest, such as cells (Gelfert, 2016). In society, to pick a few examples: models help design and run cities, manufacture cars, streamline business systems and the operation of hospitals and generate cleaner production processes.

In public policy, the past and future are often bridged by the use of computational models, such as those of the climate, of the economy, or of the spread of an infectious disease. Here, models play a particularly important role because they not only create robust evidence, they also do it in a way that improves the quality of debate around that evidence. This is partly because they act as vehicles to convene groups: those who supply the data and the expertise, those who must inform and make decisions about the questions for which the model should be developed and to which it should be applied, those who make judgements about the assumptions that underpin the model, and those affected by the model's outcomes.

This convening function means a wide group of stakeholders, with different forms of expertise can develop, challenge or use the model. However, as discussed in Chap. 1, these are often models of systems or, more frequently, parts of systems, which are different from the systems the policy-maker would actually like to be able to model.

In 2013 the McPherson Review of the quality assurance of models in use in the UK national government found over 500 models with a significant influence on policy (McPherson, 2013). They ranged from basic Excel spreadsheets constructed by inhouse analysts to the models of global climate created by international teams and subject to extensive peer review through the Intergovernmental Panel on Climate Change. The Review, set up after the errors that led to a major economic and political failure in the Department for Transport's review of the West Coast rail franchise, highlighted several risks with the use of models. In particular, it can be very tempting and easy for a model created for one purpose to be stretched to apply to another to which it appears to fit but for which, in practice, it is less well suited and may even be misleading. Following the McPherson Review, the government introduced guidance aimed primarily at ensuring proportionate external review and challenge of business critical models.

The scale and scope of the class of models that rely on computation is rapidly increasing and will continue to do so. Increasing volumes of data, smarter algorithms and cheaper computing power combine to increase the range of situations to which computational models can be applied and the reliability with which they can be used. They enable better understanding of complex systems, and create insight into the behaviours and possible futures of those systems.

Models are powerful, but can also be seductive and misleading. The basic lessons to ensure they are used wisely are to ensure that the data is good, that the policy client is engaged with the modelling experts throughout and that the assumptions behind and limitations of the model are well understood both by its designers and those who may act on its findings.

Machine learning techniques and other forms of data science are further extending the range of circumstances in which scientists and policy makers could and should use modelling and they have the potential to open up many new areas of policy insight, evidence and creativity. They also create some new challenges. One is that they enhance the range of circumstances under which seductively powerful findings, particularly when presented visually, can persuade decision-makers or publics to believe in or act on relationships that are not sufficiently well understood. Poorly applied, they can increase the risk of confusion between correlation and causation, or reinforce the effects of unexamined bias in the data through statistical stereotyping.

The resolutions will be different in different circumstances. Where the technology is perceived as more mundane and the systems around it are trusted, societies and individuals are entirely comfortable relying on things that have been proven to work well but which they don't understand. Most individuals don't understand why an aeroplane flies safely and aspirin was used for pain relief for many years before there was any good description of how it worked.

Meanwhile, the options for policy-makers to incorporate machine learning systems into accountable decision-making are likely to increase in the future. There is much research into developing machine learning systems that also provide explanations, interpretations or accounts of how they derived their findings. However in some circumstances policy-makers may have to decide to trade-off precision for accountability, as a system that can be interpreted may be less accurate than one which cannot (National Academy of Sciences & The Royal Society, 2017).

Less often discussed is the extent to which qualitative mental models of how the world works form the basis of policy decisions and public debate: models of families, interpersonal relations, communities and nations. In addition, for decision-makers, the question may not be about how they believe the world works, but about what they believe about how the world *should* work, and how far to impose that belief on what the model tells them.

This conundrum is thrown into sharp relief by machine learning, with the widely discussed problems of statistical stereotyping. If a system only makes deductions about the future by learning from historic data it will project historic patterns forward. If the data is biased (in the scientific sense of being skewed, rather than prejudiced) then the projections will be biased too. In these instances the new uses of historic data are forcing or helping stakeholders to consider how their desired models of the future differ from the realities of the past, what the bases for those differences are and how they can more explicitly design policies and practices that embed their chosen values and aspirations as well as drawing on the best data available.

WHY FUTURES WORK IS ESSENTIAL

Whereas discussions of modelling in policy tend to assume a scientific basis and tone, work on "futures" is often considered dangerously speculative. To caricature: to talk explicitly about the future quite often leads to the speaker being accused of trying to predict or forecast what cannot be predicted or forecast, of trying to control outcomes that cannot be controlled, and of general over-reaching and silliness.

It is essential to overcome this caricaturing. While unreasoned, ill-founded or over confident assertions about the future are at best a waste of time and at worst dangerous, to use fear of this kind of over-reaching to avoid being thoughtful and explicit where we can is also dangerous. Informed anticipation is the basis of resilience to future risks and of the motivations to create new knowledge, ideas, and actions.

It is worth starting from the position "Prediction is very difficult, especially about the future"[1] and then going on to recognise that humans are all always making assumptions about the future, because it is impossible not to. The question is whether we examine them or not, and what we do today in anticipation of those potential futures.

Without examination, individuals are systematically misguided when they think about the future. All the cognitive biases and mental shortcuts

---

[1] Multiple similar versions of this statement are attributed to people such as Niels Bohr, Yogi Berra and Samuel Goldwyn.

come in to play. For example, people find specific stories about the future more likely to be true than general possibilities, even when this is logically impossible: surveys show people putting put the chance of a major earthquake in, say, California to be higher than the chance of any form of future major natural disaster, even though the latter group includes earthquakes. People tend to use the near past as a default predictor of the future. Survey respondents' estimates of the frequency of extreme weather events correlate with their personal experience of such events; so something experienced recently is more likely to be expected to happen again soon (Ipsos MORI Research Institute, 2013).

On the other hand, once anticipating change, there are important circumstances under which some people may instinctively over-estimate the effects in the short term, and under-estimate them in the longer term, perhaps related to the fact that the longer term changes are likely to be harder to imagine. A former head of the UK Prime Minister's Strategy Unit referred to this as an optical distortion, in which governments themselves may overestimate the impact of their short term measures and underestimate what can be achieved longer term (Mulgan, 2005). The same is often said of the nature of wider public debate about emerging technologies. However the pattern is complicated by the tendency of utopian and dystopian language to be adopted in parallel in some parts of the debate.

For public policy, the question is then about how to examine assumptions about the future explicitly and in ways that best inform democratic decision-making. Some forms of futures work operate simply to destabilise or make visible pre-existing assumptions about what the future will be like. It may also be helpful to consider rigorous futures work not as prediction, but as a form of engaging with uncertainty, of understanding some of the consequences of complexity, of integrating multiple disciplines' perspectives, and of making systems analysis intelligible. The boundary between well-informed futures work and the practice of using scientific evidence in decisions is not absolute.

A report by the innovation foundation, Nesta, "Don't stop thinking about tomorrow" (Bland & Westlake, 2013) proposes three modes of futures thinking. The first is forecasting, as in forecasting the weather. Data enabled and computational methods based on accepted models outline well-defined aspects of the future and allow estimates of the uncertainty about a range of plausible values for those futures.

The second is foresight, which Nesta defines as the process of developing distinguishable versions of the future, with a sense of the drivers that will shape them, but with no attempt to estimate the relative likelihood of different versions. These versions are often presented as scenarios; complete, internally coherent descriptions of a system. They typically represent attempts to describe possible future states of complex systems such as flood risk or cities. Foresight capabilities help governments address systemic challenges and opportunities. They are, for example, a well-established part of national approaches to strategic government in Singapore. Foresight is also one of a suite of tools which include behavioural insights, horizon scanning and "policy laboratories" provided by the EU's Joint Research Centre, for enabling policy-makers to engage with scientific evidence.

Nesta defines the third mode as fiction. It argues that fictional stories about the future inform innovation, express imagination and enable the process of sharing desires and fears about possible futures in order to help shape it (Borup, Brown, Konrad, & Van Lente, 2006). For entrepreneurs from Elon Musk to Martha Lane-Fox inspirational stories about futures they consider desirable create the irrational persistence necessary to bring about some forms of transformative change. As with much about the future, fear and longing and imagined dystopia and utopia, are often inseparable (Atwood, 2011). The programme to go to the Earth's moon was driven by both, as are current aspirations around AI, terraforming and planetary travel. Other uses of stories of the future are discussed later in this chapter.

Uses of Foresight and Anticipation

The term "foresight", like "futures" and the emerging term "anticipation", is defined and used differently by different groups. However key features of foresight are illustrated by the UK Foresight programme.[2] From 2002 onwards, the UK government invested significantly in a central strategic futures capability. This version of Foresight was intended primarily to inform Ministers, but to do so in ways that supported longer term decision-making by combining futures work with the creation of synthesised evidence relevant to major public policy topics. In addition to adopting specialist futures techniques the programme's overall approaches reflected then current approaches to strategy. It drew heavily on the

---

[2] Material relating to all Foresight projects is available through the UK Government website, gov.uk.

learning from corporate strategic approaches to uncertainty; the origins of many late twentieth century approaches to futures thinking were in Shell's corporate scenarios work, which was attributed with helping the company be more resilient than its competitors, to the 1974 oil price shock.

The Foresight programme was designed and overseen by successive UK Government Chief Scientific Advisers (GCSAs), starting with Sir David King. It experimented with different futures techniques, and evolved over time to suit different strategic objectives. This led to a body of practice that has some enduring lessons on approaches to scientific knowledge, evidence and the future.[3]

Between 2002 and 2016 Foresight conducted over 20 projects. One of the earliest, Future Flooding (Government Office for Science, 2004) was on flood and coastal erosion risks in the UK. Part of the project's authority derived from its combination of computational modelling with scenario-based futures techniques looking out to 2080. The conceptual model was of "source, pathway and receptor" and based on 17 main variables, which ranged from rainfall to land use patterns. The quantitative basis was historic data on flood damage available in England and Wales collected on a 10 km square grid. The expert team combined this with the Hadley Centre's models for global climate change, to create quantified scenarios to 2080.

The scenarios themselves were derived by classic futures techniques in which groups of experts and practitioners discuss and develop the key drivers of future uncertainty, assign ranges of values to them and identify the most significant or the ones of most interest. They then develop internally consistent and plausible narratives of the future for a limited number of combinations of different values of the key drivers. The most common approach is to take the two drivers of greatest interest, and develop four scenarios on a 2 by 2 matrix for high and low settings of each driver. For flooding, this meant the production of scenarios that considered high and low settings for climate change impacts and for economic growth. The model was then used to test plausible baskets of responses such as building flood defences, or adopting different patterns of land use for housing and flood management.

This work, which deliberately straddled and pushed the boundaries of modelling and futures, showed flood risk going up under all plausible

---

[3] In his book, Science and Government, C.P. Snow describes Foresight as "not quite knowledge ... more an expectation of knowledge to come" (Snow, 1961).

assumptions about the future. That message, of itself, contributed to the political and public debate on climate change that was part of the context for the UK's international political leadership in setting statutory budgets for future carbon emissions in the Climate Change Act 2008.

Estimates of total future risk were, of course, extremely uncertain. Estimates of the increase in annual economic damage by 2080 varied from £1.5bn to £20bn, with most of the difference being due to differences in assumptions of economic growth. Nevertheless, government's immediate response was to increase investment in flood management by £350m, with HM Treasury citing the report as evidence informing that decision.

In this project, the model enabled audiences to consider uncertainty, and to discuss and communicate it with numbers, with images and with words. As discussed in Chap. 2, numbers are always compelling, but images are too. Show the maps of future risk to an ordinary person and they look for the 10 km square their house is in, to see whether the risk is shown as red or green. Show them to a Minister or other Member of Parliament, and they look for their constituency. The general becomes specific in an instant even, according to the then GCSA, at the Cabinet table.

The existence of underpinning quantified modelling meant the approach was transferable to other geographies provided there was sufficient data to inform the model. After publication of the original report some of the Foresight team of experts subsequently worked with Chinese authorities to consider flood risk in the Taihu basin (Shanghai). The approach was also used by the US Army Corps of Engineers and by the UK's Thames 2100 project to inform decisions on the future of the Thames barrage.

In addition to creating transferable models for conceiving possible futures, Foresight projects split the elements of reasoning about future uncertainties in ways intended to support public debate over time. In all cases, the starting point of the project was the creation of a synthesised evidence base drawing on the largest possible number of relevant disciplines. The project considering future trajectories for Brain Science, Addiction and Drugs, for example, drew explicitly on more than 20 disciplines and the Mental Capital project generated over 70 supporting papers.

Creating these reviews of the evidence within disciplines but for a shared project, resulted in novel inter-disciplinary conversations. The conversations sometimes created relationships resulting in new research agendas and the reviews themselves were used, well after the lifetime of each project, by researchers within and across disciplines. The reviews also resulted in a body of evidence that could be considered separately from the more speculative futures work and interpreted differently by successive decision-makers with different political priorities. From the perspective of the scientist in Honest Broker mode, provided all the conclusions were informed by the evidence, this was success.

PLAYFUL MODELS AND SERIOUS GAMES

Models, whether used as engines for futures work or as cameras to explore near term uncertainties, act as vehicles for convening debate. While this is true within the technical discussions, it is also true with public discussions. Giving away control of the model's parameters takes the possibilities of getting reasoned debate about uncertainty a step further.

In addition to using their model to inform government decision-making directly, the Foresight team used it to create a software-based simulation game called FloodRanger (Discovery Software). The game was based on an anonymised English region. The player becomes the political leader and makes investment decisions for the locality such as to support flood risk management, economic growth and housing. Then the game steps forward a decade, assigning climate impacts based on global climate modelling. Depending how those impacts affect a basket of local outcomes, the player is voted in or out. The aim is to stay in power for a century.

FloodRanger complemented the traditional report which, as is the way of these things, weighed in at over a kilogramme in hard copy, including the annexes. The game was subsequently used by the Environment Agency as part of the public dialogue at local level.

Handing over control of the model's parameters gives the player, whether a national decision-maker or interested citizen, an opportunity better to grasp the uncertainties about future outcomes. However, it goes further, in that repeated playing allows for a more intuitive feel for the characteristics of a highly nonlinear system. For example, the same investment choices at the start lead to different outcomes in different runs of the model, allowing the player more fully to inhabit the realities of making decisions under these types of uncertainty.

For those prepared to make bold simplifying assumptions, playful models can act as powerful illuminators of potential futures. The 2050 Calculator (Department for Business, Energy and Industrial Strategy,

2013) originally introduced by a Chief Scientific Adviser is, in scientific terms, a simple spreadsheet model with a reasonably user-friendly front end. It allows anyone with an interest to test out plausible pathways to the UK's 2050 carbon targets. What it does not do is allow a user to design what they would like to see, where to do so would violate essential constraints. This forces users to confront policy trade-offs that they might otherwise choose to avoid.

However powerful such computational models can be, in most Foresight projects it was simply not possible to create or use them plausibly. The systems were incapable of sufficient definition, such as those concerned with the future of cybersecurity, or too large, such as those concerned with obesities. In these and other projects, the teams typically modelled sub elements in order to inform overall findings presented in a mix of qualitative and quantitative narrative.

The Foresight project on Cybertrust and Crime Prevention developed qualitative scenarios, each an internally consistent account of a plausible mid-term future set notionally at 2018 as seen from 2008. The qualitative scenarios were used in workshops with various stakeholders including, unusually, a government Minister.

In one instance, the scenarios were applied to a then immediate policy question, that of whether to introduce electronic tagging to enable some offenders to be kept out of prison but still at a known location. A concern at the time was that knowing where someone was at all times was so unusual that it would of itself be stigmatising in a damaging way. However, in the imaginative act of inhabiting a future in which technologies not then available were widespread, it became clear that knowing where someone was located was likely to become much more commonplace and hence that it might not of itself be uniquely stigmatising to offenders.

Serious gaming is another way in to the challenge of developing potential futures and of enabling decision-makers to inhabit them. Role-play, red-teaming in which one group plays the part of the enemy, and even board games are all used to answer questions about the present, and—like many techniques that are ultimately part of strategic thinking—are employed in the military and business. It is a relatively small shift to move from thinking about alternative futures that might result from decisions taken today and enacted in what is otherwise today's world, to thinking about alternative futures in plausible worlds of tomorrow.

Some Roles of Narrative

Narrative[4] plays many roles, amongst which are its capacity to shape imagination, to influence which evidence is noticed, by whom, and with what effect, and to be essential to the process of reasoning about, making and defending public policy decisions. Knowledge about which narratives are influential in which groups can itself provide a form of evidence (for a flavour of the different potential approaches see, for example Morgan (2017), Nature (2018), Le Guin (1989)).

The notion of "story" is taken from narrative theory and used to describe the patterns that structure human experience and which are essential for initiating, performing and remembering (Turner, 1996). This process extends from tiny events such as grasping an object, through to becoming the building blocks of cognitive activity and ultimately into forming the everyday mind, which is, in this sense, a literary mind.

Because of its deep links to the organisation of memory and communication and its integration of cognition and emotion, narrative affects decision-making. The public decision-maker is using both cognition and emotion to make the decision and must use both of them when explaining, defending or persuading others to accept it. In any given policy moment they will want to adopt a causal story: who is responsible and who has agency? They will also create and use narratives with emotional effect: what will the outcomes *feel* like for the citizen?

By their very nature, narratives not only form the basis of individual thought, they are also usually easily remembered and shared and so shape the public framing of issues and creation of shared worldviews. Studies on how narratives point up the saliency of particular types of evidence show that. For example social scientists' work on phenomena such as social amplification (Pidgeon & Barnett, 2013) and "elite cues" (Carvalho & Burgess, 2005) discusses the conditions under which particular types of evidence and framings come to the fore in public debate.

Narratives direct attention towards or away from particular types of evidence, influence which accounts spread most rapidly between individuals and groups and affect or describe the extent to which particular groups

---

[4] As always, the use of related terms varies across disciplines and authors. Here narrative is used to mean a particular account, whereas a story is a sequence of events.

trust it or feel they have an interest in and agency as a result of, its implications (Hustvedt, 2016). For example, stories about the agency of and relationships between parents, patients and professionals affect public and private decisions on vaccination, or the safe sharing of health data. Literary fiction in the sense of the novels of Gaskell, Dickens or Hugo is also associated with directing public attention at aspects of the historical present and so helping create the conditions for change.

At the same time, narratives inform the motivations and identities of researchers, the shared interests and assumptions within communities of researchers, the shape of research questions that are asked and funded and the extent to which new evidence or ideas are picked up more widely. The practices of research about endangered species may begin with studies of artefacts such as photographs and film but they inform the creation of the Red Lists of endangered species and the laws that affect them, while narratives of the Anthropocene—of loss and anticipation in connection to humans' relations with plants, animals and other living and inanimate aspects of the Earth—influence both today's academic and public directions (Heise, 2016).

Reflecting on the role of narrative highlights the linked issue of embodiment. Narratives require narrators, agents and objects. As discussed in Chap. 1 the material and, in particular, the human, humanoid or anthropomorphised, accounts are typically more compelling, picked up and used, in contexts from discussion with a Minister to Twitter. A story about providing a new therapy to cure a sick child is superficially more compelling than considering distributed and messy interventions in a public health area such as obesity. Badgers or bees become the embodiment of public reasoning about biodiversity or food safety.

Every significant public policy question is a question about the behaviour of a complex system and these comparisons point up a fundamental challenge of engaging with evidence about systems. It is almost impossible to take a systems approach to anything and make it sound immediately compelling to a generalist audience. So narrative and embodiment, like models, are in this context at the very least valuable tools to think and communicate with (Currie & Sterelny, 2017).

The challenge is to reflect on narrative (how the stories are being told) as well what the stories are, in order to inform and enable well-founded debate and public reasoning. Narratives' energy and insights may draw attention to aspects of a complex debate that are also systemically significant, or to aspects that might otherwise inadvertently be overlooked. Or it may be that they are helpfully illustrating a completely different framing of the issue at hand, such as one based wholly on matters of principle or ethics, and so pointing importantly to particular lenses and framing beyond the scientific evidence, as discussed in Chap. 4.

Discussion of narrative's roles in the tasks of enabling cognitive and sentimental engagement with complexity and multiple framings is slippery territory, not least because there is no avoiding the reality that non-fictional accounts always carry emotional impact. The boundary between fiction and non-fiction, and engagements with rationality and emotion, has never been as fixed as some would like it to be.

NARRATIVES AND FUTURES

Decisions, and debates about their possible consequences, require the creation of options imagined to take place in the future. Policy debates often assume a relatively straightforward imposition of various options onto a landscape that is otherwise unchanged, with no second order or systemic consequences. That may be entirely proportionate. But as the significance of the decision gets bigger, so does the need to consider these further consequences. They may be harder to engage with either because they require the modelling of more complex systems, or because they are further into the future and therefore the number of plausible potential paths has grown too. Either way, narrative continues to carry its roles of enabling emotional engagement with consequences, finding new cognitive ways into complexity and, in the case of extreme speculation, creating rapid and accessible forms of anticipation (Raven & Elahi, 2015).

As the Foresight Flooding and Cybertrust scenario examples illustrate, decision-makers can engage through narrative in futures work. This helps ensure public reasoning about the future is better informed than it would have been if relying on instinctive notions that tomorrow will be much like today.

There is increasing interest in the roles of Speculative Fiction (SF), including alternative pasts, presents and futures (Bassett, Steinmueller, & Voss, 2013). These stories appear to warn and inspire. They act as thought experiments about the implications of specific new technologies or social arrangements. They shape language and enable new ways to prototype artefacts. They reinforce the identities of some scientific sub-groups and inform what questions get asked or researched (Dillon & Schaffer-Goddard, n.d.). It is certainly the case that it helps to have some knowledge of SF if engaging with a group of, say, computer scientists.

Like stories, when used carefully, material objects can act as ways to develop new insights. In carrying out conversations about aspirations for the culture of research in 2035, participants in a Royal Society programme generated a Museum of Extraordinary Objects. For example, the Noble Award could only be awarded to teams, rather than to individual scientists, and the Heroic Bracelets memorialised failure (The Royal Society, 2018). This process of imagining and making fictional objects also enables new forms of public and other engagement, discussed further in Chap. 4.

In the end, narrative fiction is a powerful way of anticipating long term human futures. Understanding more about its operation is particularly important with respect to public debates about aspects of Artificial Intelligence and robotics. Both come freighted with stories from the past, Western records of them starting with the Oracles and mechanical people of Ancient Greece and Rome. Some stories are visible consistently through to today's fictional and non-fictional accounts: that of the roles of tools or slaves; of the Apprentice that becomes wiser than their Master; or of the anticipation of a life of luxury and hedonism that is also dangerous or false.

Fictional and non-fictional accounts of AI and robotics are also strongly influenced by the relative ease of discussing embodiment rather than systemic effects. It is easier and more compelling to imagine humanoid robots than to consider the evolution of the consequences for business models, organisations and labour, of the thousands of robots already in use in production processes. For Autonomous Vehicles it is easier to imagine and discuss whether future vehicles might increase or decrease traffic deaths, than to begin to consider the impact their significant uptake might have on the mobility of the elderly and disabled, on increasing obesity, or on the evolution of the layout of cities—even though the twentieth century shows the latter two to have been vastly influenced by the initial introduction of the car.

References to humanoid robots such as the Terminator (Cameron, 1984) are used in non-fictional accounts to draw attention to AI, but bring with them associations of Artificial General Intelligence which is still many years off. Their use may therefore distort expectations about near term technological trends and distract from, say, the discussion about embedded forms of intelligence in buildings, or cities.

In the room together discussing AI, the Minister and the Nobel Laureate may have to choose whether to dismiss current references to the Terminator story about a nearly-unstoppable malign robot as simply a communicator's potentially lazy (albeit sometimes helpful) attention-grabber; to pause to consider whether it is distorting public debate by making human-level intelligence seem nearer that it probably is; or to decide to consider it a metaphor for some public concerns about potentially unstoppable malign forces of technology, industry or government. They will almost certainly do the former, but their advisers should not always be so dismissive.

The task of looking further forward is also well informed by looking further back. Anticipatory histories help destabilise the unexamined default assumption that the near future will be like the recent past (de Silvey, Naylor, & Sackett, 2011). History, like many other areas of the humanities also helps provide important counterweight to some of the risks created by unreflective adoption of dominant narratives, and by narrow focus on narratives of risk, not least by helping enlarge and inform the space for thinking (Edgerton, 2006). So for example, in discussions about the implications of machine learning's "automated decision-making" in commercial organisations, the historian's account of how the introduction of the mathematical technology of standardised financial accounting contributed to the enablement of fully globalised commercial organisations raised the sights from the personal to the systemic and global.

NARRATIVE UNDERSTUDIED?

Over more than 15 years the UK Foresight programme engaged formally with narrative only twice.[5] One was to include a review of autobiographical accounts of addiction over the twentieth century as part of the project on Brain Science, Addiction and Drugs. Those present felt a frisson of excitement when the authors of the 20 or so reviews "revealed" their key findings to each other: and the neuroscientist and the historian of medicine described—from wholly different and independent perspectives—the

---

[5] In an early and rarely cited project exploring potential future uses of the electromagnetic spectrum the team directly commissioned Kate Mosse, before she became a best-selling author, to write four short pieces of fiction. These were presented as part of the final package of project outputs, alongside technical roadmaps. There is no report as to what happened as a result.

importance of the first powerful experience of addictive drug use, for the subsequent pathway of the addict.

This account illustrates a vastly under-explored area of the practice of evidence in government. Compared to the time, money and effort put into, say, developing computational modelling for decision-support, the amount invested in rigorous thinking about narratives is tiny. This may reflect an underlying nervousness about exploring some aspects of the bases of individual decision-making at the hearts of both democracy and science, for different reasons. Discussion of narrative, inevitably and explicitly, brings in to play emotion and subjectivity. Researchers and commentators examine the professional and public reasons for a Minister or a government making a decision, but typically not the private and personal ones unless they relate to wrongdoing. It also destabilises the passive third person singular account of the scientific paper, and steps out into the territory of philosophy of science in which, for example, "all models are fiction".

Those concerned with public reasoning and science are only beginning to be able to ask and answer questions about how new narratives form or existing narratives become salient to debate; how they are amplified or diminished; and when and how they affect outcomes. And yet, over the next few years, the impact both of available data and of smarter software alongside scholarly insight, will create new opportunities. It should be increasingly possible to find out more about the way narratives connect individual mental models of the future with the observational and quantitative evidence provided in the professional advisory environment; the issues researchers choose to research; and the social contexts that do or don't lead to uptake and regulation.

As a single illustration of the kinds of wholly new forms of approach that might become possible, consider the example of a review of English language film and TV transcripts over the last decade. This showed that "control" features as a significant part of stories in which AI plays a part, and more strongly than for other technologies that might also be considered novel and risky (Recchia, 2017). In the Royal Society's public dialogue on Machine Learning, the issue of control similarly emerged as significant (The Royal Society, 2017). Finding out more about such relationships will fill a major area of strategic ignorance about how aspects of anticipation flow between public and private debate; fiction and non-fiction accounts; scientific, literary and popular spaces.

REFERENCES

Atwood, M. (2011). *In other worlds: SF and the human imagination.* Hachette Digital.

Bassett, C., Steinmueller, E., & Voss, G. (2013). Better made up: The mutual influence of science fiction and innovation. *Nesta.* Retrieved April 14, 2018.

Bland, J., & Westlake, S. (2013). Don't stop thinking about tomorrow. *Nesta.*

Borup, M., Brown, N., Konrad, K., & Van Lente, H. (2006). The sociology of expectations in science and technology. *Technology Analysis and Strategic Management, 18*(3/4), 285–298.

Calder, M., Craig, C., Culley, D., de Cani, R., Donnelly, C. A., Douglas, R., et al. (2018). Computational modelling for decision-making: Where, why, what, who and how. *Royal Society Open Science, 5*(6), 172096.

Cameron, J. (Director). (1984). *The Terminator* [Motion Picture].

Carvalho, A., & Burgess, J. (2005). Cultural circuits of climate change in U.K. broadsheet newspapers, 1985–2003. *Risk Analysis, 25*(6).

Currie, A. M. (2017). From Models-as-Fictions to Models-as-Tools. *Ergo: An Open Access Journal of Philosophy, 4.*

Currie, A., & Sterelny, K. (2017, April). In defence of story-telling. *Studies in History & Philosophy of Science, 62,* 14–21.

de Silvey, C., Naylor, S., & Sackett, C. (2011). *Anticipatory history.* Uniformbooks.

Department for Business, Energy and Industrial Strategy. (2013, January 22). 2050 Pathways. *Gov.uk.* Retrieved April 15, 2018, from https://www.gov.uk/guidance/2050-pathways-analysis

Dillon, S., & Schaffer-Goddard, J. (n.d.). What artificial intelligence researchers read. Forthcoming.

Discovery Software. (n.d.). *FloodRanger.* Discovery Software Ltd. Retrieved April 22, 2018, from www.discoverysoftware.co.uk

Edgerton, D. (2006). *The shock of the old: Technology and global history since 1900.* Oxford University Press.

Gelfert, A. (2016). *How to do science with models: A philosophical primer.* SpringerBriefs in Philosophy.

Government Office for Science. (2004). *Future flooding.* Office of Science and Technology. Retrieved April 14, 2018, from https://assets.publishing.service.gov.uk

Government Office for Science. (2018). *Computational modelling: Technological futures.* Government Office for Science. Retrieved April 2013, 2018, from https://assets.publishing.service.gov.uk

Heise, U. K. (2016). *Imagining extinction: The cultural meanings of endangered species.* University of Chicago Press.

Hustvedt, S. (2016). *A woman looking at men looking at women.* Simon & Schuster.

Ipsos MORI Research Institute. (2013). *PREPARE – Climate risk acceptabilty. Findings from a series of deliberative workshops and online survey.* Department for Food, Agriculture and Rural Affairs, Defra. Retrieved April 22, 2018, from https://www.ipsos.com

Le Guin, U. K. (1989). *Dancing at the edge of the world*. Grove Press.

McPherson, N. (2013). *Review of quality assurance of government models*. HM Treasury. Retrieved April 14, 2018, from https://www.gov.uk/government/publications

Morgan, M. (2017). Narrative science and narrative knowing. *Studies in History and Philosophy of Science Part A, 62*, 1–5.

Mulgan, G. (2005, April 23). My time in the engine room. *The Guardian*. Retrieved April 14, 2018, from https://www.theguardian.com

National Academy of Sciences; The Royal Society. (2017). *The frontiers of machine learning*. National Academy of Sciences; The Royal Society. Retrieved April 22, 2018, from www.nasonline.org

Nature. (2018, March 7). Learn to tell science stories. *Nature, 555*, 141–142.

Pidgeon, N., & Barnett, J. (2013). *Chalara and the social amplification of risk*. Department for Environment, Food and Rural Affairs.

Raven, P. G., & Elahi, S. (2015). The new narrative: Applying narratology to the shaping of futures outputs. *Futures, 74*, 49–51.

Recchia, G. (2017, May 17). *Fall and rise of AI: Computational methods for investigating cultural narratives*. Invited presentation to AI Narratives: Workshop 1, Leverhulme Centre for the Future of Intelligence and the Royal Society, Cambridge, UK.

Snow, C. P. (1961). *Science and government*. Oxford University Press.

The Royal Society. (2017). *Machine learning: The power and promise of computers that learn by example*. The Royal Society.

The Royal Society. (2018, April 14). The museum of extraordinary objects. *The Royal Society*. Retrieved from https://royalsociety.org

Turner, M. (1996). *The literary mind: The origins of thought and language*. Oxford University Press.

CHAPTER 4

How to Engage with Publics

DEMOCRACY AND SCIENCE

It is implicit in democracy that the right to have a say in a decision through voting is independent of any form of expertise and not weighted by any measure of ability. However there are modifications to that basic premise, in that democracies always do limit participation in some way, such as excluding offenders who are in prison, or setting age limits below which a citizen cannot vote.

They also limit it, for the most part as they must simply to get business done, in the multiple ways in which individual decisions do and don't get shaped and scrutinised. So in the UK most major policy decisions at some point will be scrutinised by Parliament, bringing the knowledge, expertise and influences on individuals in both Houses to bear. This points to the importance of considering how science and scientists engage with Members of the Houses. There is a major role for those interested in ensuring the quality of science in government to get involved directly with local and regional politics and nationally to engage with Members on specific issues.

Science, for its part, is also an activity in principle open to everyone. Each person is in some ways a scientist, for to be a scientist fundamentally means to be curious, to be open to observing the world as it is, and to be ready to accept the answers the world gives. However the system that creates new scientific knowledge, like the system that creates new policy, sets requirements that necessarily exclude most people from direct participation in that part of the activity. To go through peer review, intense

© The Author(s) 2019
C. Craig, *How Does Government Listen to Scientists?*,
https://doi.org/10.1007/978-3-319-96086-9_4

competition for resource, and the sometimes visceral battles required to achieve new knowledge, requires as much dedication as to get elected in anything other than a safe seat.

Where decisions are of a well-defined nature and have major effects on society or on individuals, there are in practice a wide range of constructs to enable public decision-making in complex situations. These constructs represent different arrangements for decision-making, expert advice, public deliberation and scrutiny. The major constructs are themselves the product of Parliamentary process. So, for example, the UK's Monetary Policy Committee, composed of expert economists, routinely makes specific decisions about interest rates. The Consumer and Markets Authority has expert lawyers and economists to inform its judgements about company and market behaviours. The Courts are informed by expert evidence and, in major cases, by jury deliberations.

In an area of decision-making routinely requiring a high degree of scientific expertise, the Human Fertilisation and Embryology Authority regulates fertility clinics and research involving human embryos. Over time, when considering wholly new potential genetic technologies, it has consistently followed the practice of seeking out views from multiple groups including academics, patients, practitioners and businesspeople. The general democratic process created a body with regulatory power which, in practice, deploys that power in ways informed by a wide range of voices with different types of expertise, including scientific expertise. It is considered successful by many stakeholders because it has enabled new techniques to be introduced to the benefit of specific categories of patient, with relatively low levels of controversy.

In a very different area of science and policy, the UK's Climate Change Committee was established in 2008 under the legislation that also created statutory national carbon budgets. It is a further example of a specific deliberative structure with a public profile and with capacity to commission and examine scientific evidence across multiple disciplines.

International comparisons of science advisory regimes show a very wide range of mechanisms linked, amongst other factors, to the variety of national political cultures and institutional practices (Lentsch & Weingart, 2009). However, Lentsch and others argue that there is a common factor to the success of advisory regimes, which is the extent to which their design enables them to achieve both expert and political robustness.

Many of the contested areas of science in government may come about because there is not such a structure, and the decisions become too remote. Those making them potentially face a double democratic deficit. The decisions have no mechanism for direct citizen participation, which is true of many other decisions public bodies routinely have to take. But the decision-makers are in some senses doubly removed due to the real or perceived barriers between the scientists and the publics (Jasanoff, 1990). Ask a child to draw a scientist and consistently over time surveys show that they draw a person in a lab coat. That very coat has connotations of sterile environments and closed-off laboratories. The laboratory walls may protect the conditions necessary to conduct precise and accurate experiments, and the public from potentially dangerous science, but they also inevitably inhibit the possibility of constructive exchanges between the two.

So there are times when, for questions such as how to grow GM crops with the aim of improving yields or reducing the use of pesticides and herbicides, or whether to cull badgers with the aim of reducing bovine TB infections, the available advisory and decision-making mechanisms struggle.

MANY PUBLICS

To a scientist or scholar everyone outside their own discipline is in some senses a member of the public, including other scientists, scholars and policy-makers. As discussed in Chap. 2 this helpfully means the only truly common language between disciplines is also that of members of the public. However, different publics have their own languages too, and two structures whose operations and leaders have particularly strong influences on public reasoning are the law and the market.

The Law

In a recent initiative between the Royal Society and the senior judiciary, a programme of direct engagement between scientists and judges is helping to identify ways of ensuring that scientific disputes in court focus only where there is reasonable doubt about the evidence. The programme has considered topics such as memory, pain, machine learning, and DNA. One of its outputs is a suite of judicial primers providing a synthesis of the evidence relevant to a particular topic. The purpose of synthesis here is the same as that for synthesis provided to policy-makers and discussed in Chap. 1: to reduce unnecessary disputes and so to enable the system to focus on the aspects of debate, reasoning or judgement that matter most, by making the current best scientific evidence more easily available and accessible. The president of the UK Supreme Court distinguished four types of primer or synthesis: those dealing with the science of forensics, such as the uses of DNA or gait analysis; those about "pure" science, such as the characteristics of machine learning algorithms; those about good practice in areas such as specific medical techniques; and those about the scientific method, such as understanding probability and the use of statistics (Neuberger, 2016).

Looking more broadly at the interplay between science and the formulation of law exposes fundamental questions about agency. As a consequence of new science and technology, lawmakers debate whether robots should have rights and the advent of machine learning and networked systems unsettle the nature of the legal role of corporations and national armed forces. The question of the implications of AI and automation for the future distributions of work, income and wealth depends in part on the relative power of those who provide labour and those who provide capital.

More practically and immediately, scientific evidence has a major part to play in enforcing existing law. Activist lawyers at ClientEarth pursue cases on behalf of the planet. In doing so, ClientEarth uses scientific evidence to direct attention not in order to change the policy, but to ensure the implementation of existing polices. In areas such as air quality, it has successfully brought prosecutions in the US and EU.

Where commitments are non-binding and there is no dedicated statutory framework, scientific evidence also has a key role. The definition and measurement of the UN Sustainable Development Goals are informed by science and scholarship. The question of how they are measured has been integral to their shaping from the start, because measurement will be necessary to motivate and demonstrate progress. Science and scientists have an important role to play in ensuring the measures are practicable, meaningful and visible.

The Market

Businesses are themselves generating vast amounts of scientific knowledge and businesses and businesspeople play significant roles in informing public decision-making. In the context of science and government they are therefore another source of the many "publics". This section highlights two issues relevant to the pursuit of all policy goals, not only those concerned directly with economic growth.

The first is the way in which leading edge science informs aspects of markets which in turn influence non-economic policy outcomes. Scientific evidence informs the development of many of the regulations and standards that shape products and services and their environmental and human impacts throughout their lifecycles. Patterns of investment and insurance will respond more precisely to non-economic indicators if that evidence is presented skilfully. For example, the insurance and re-insurance sectors invest extensively in the modelling of environmental, physical and other natural disasters, in order better to account for their potential future impacts. Such accounting has the effect of hardwiring scientific knowledge about the planet in to market incentive structures.

The second is that, from the craft skill perspective, engaging businesses in providing scientific evidence for policy decisions presents a straightforward but insoluble dilemma. In many policy areas to exclude business-people or evidence provided by businesses would substantially reduce the knowledge available and therefore the quality of the debate and the decision. To include it risks significantly reducing the confidence that wider publics will have in the decision. Public surveys consistently report high levels of relative trust in scientists in universities compared to those in industry (British Science Association, 2015).

In most cases the best, and perhaps only, response is to place immense care in designing the deliberation and decision-making processes. These, proportionate with the outcomes at stake, should enable access and transparency in line with the principles that underpin high quality evidence synthesis outlined in Chap. 1.

LENSES

One of the challenges of linking theory and practice in the area of science in government is that the language, both academic and popular, concerning some key concepts is still unsettled. (Susan Owens refers to a "large and somewhat unruly literature" about the interface between knowledge and policy (Owens, 2015)). There are many conceptual frameworks and theoretical terms seeking to tackle similar underlying concepts but each doing it slightly differently. Each senior science adviser typically has their own preferred language too.

A particularly important challenge is that of establishing a common description for the task of defining the focus and scope or boundaries of the issue under debate, delineating the principal aspects of it that should be considered, and acknowledging the different potential stakeholder perspectives on each of these things. One possible way of differentiating between overlapping terms commonly in play is as follows. "System" can be used to mean the "scientific" definition of the boundary and major elements of the issue, including questions of agency within the system. A model is typically a causal framework for understanding part or all of the system and, in particular, for enabling predictions about its behaviour. "Framing" is an account of the issue that foregrounds worldviews, narrative, values and emotions. In doing so, it will implicitly define or assume the existence of systems.

However, from the perspective of public debate the term "lenses" may also be useful. This has something akin to the "policy image" of writers such as Paul Cairney but the word was used frequently by Sir Mark Walport as UK Government Chief Scientific Adviser (GCSA), possibly betraying his disciplinary roots in medicine. As a metaphor it helpfully conveys a degree of agency: the user can adjust or change their lens. Two people can apply the same type of lens, although they bring to it their own eyes. The lens metaphor conveys that it is possible for anyone involved in a policy question to go some way towards being able to recognise and distinguish between different aspects of a complex issue and do so in a manner that is helpful for providing evidence and for debating outcomes. It reflects pragmatic acceptance of the need to make choices in providing evidence or advice, in the knowledge that there is no single approach that will be right and complete, from every available perspective.

Taking the example of hydraulic fracturing for gas: this has at least three significant lenses. These are the risks of earthquakes and groundwater contamination, the impact on energy prices and global carbon emissions, and the impact on local amenity (countryside and roads). They are all important, but resolving them requires different types of evidence and engagement with overlapping but different stakeholder groups. Acknowledging the existence of the different lenses enables better-informed debates and parts of the issue to be moved forward faster.

When stories of "earthquakes in Blackpool" began to be reported in the media, the then GCSA commissioned the Royal Society and Royal Academy of Engineering to review the health, safety and environmental risks associated with fracking (The Royal Society & The Royal Academy of Engineering, 2012). The scope of the request prompted considerable discussion about what evidence to provide. For example, some scientists initially took the view that it was more important first to provide evidence on the implications of fracking for future carbon emissions and climate change. However, keeping the scope restricted to the single lens of risk meant the report could be delivered in a few months, in time to inform scheduled policy and regulatory decisions. The relative speed with which it was completed may also have encouraged public debate on the other lenses, by helping to avoid the question of earthquake risk becoming a lightning rod and dominating the whole.

FORMS OF SCIENTIFIC EXPERTISE

Just as each of us is a member of the public, so we are all experts. A Foresight study on the future of identity described how social and technological trends create opposing effects at the same time. Each individual has aspects of identity that they foreground in different contexts, such as campaigner, worker, parent or sibling. Compared to their predecessors today's generations have more options to foreground identities that were previously too rare to be nurtured in groups or networks at scale. These might be identities such as that of being a patient living with a rare medical condition or a passionate practitioner of a rare craft or sport. However the possibility of keeping some different aspects of a personal identity entirely separate, previously not much noticed because it required little active effort, has almost completely collapsed. My personal interest in Speculative Fiction is easily visible alongside my professional interests in physical sciences and policy, to anyone who cares enough to look.

Being expert in lace-making, family law or aikido does not help with decision-making on, say, the future of hydraulic fracturing for shale gas. Those same individuals may, however, have any one of a number of expertises that do have a bearing. Harry Collins distinguishes three principal types of "scientific expertise" (Collins, 2014).

Ubiquitous expertise is that which each individual acquires from growing up in their society. It helps them navigate their world.

Specialist expertise itself comes in two forms. The first derives directly from experience. That experience may be in the form of an apprenticeship: whether leading to the set of specialist skills necessary for a formally specified job, or a defined role such as parenting. It may have been acquired by learning how to live with a chronic illness, or by becoming expert in a leisure activity. It may have been learned by working closely with the specialist expert group. People in this category therefore gain the ability to understand how to "read" the expert knowledge of the scientific specialists.

It includes the brokers who operate at the interface of science and policy not as scientists or as generalist policy makers, but as experts in knowledge exchange.

The second form of specialist expertise represents those who acquire knowledge, but without having the context that enables them to use it to its full extent. Reading, watching or listening to news coverage of scientific research creates this form of expertise.

The third type is meta-expertise. Again, this comes in two forms: one is expertise about other people. This may be general, such as most humans' abilities to move from one town to another and navigate both the physical and social environments fairly well including, say, working out who is a police officer or who is a sex worker. Or it may be more specific, as for a group of scientists working closely together within a discipline or research group.

The second form of meta-expertise is the expertise of judging expertise, which itself can be done well or badly. So an individual may employ an architect to make judgements about the work of the builder. In synthesising scientific evidence to provide it in a form accessible to a policy-maker it is frequently the case that one expert will be required to make judgements about the quality of claims by another. If both experts are from the same discipline, this is a form of the peer review that is the underpinning of science as a whole.

More frequently an expert in broker mode, such as a Chief Scientific Adviser, will have to use expertise gathered in one domain or discipline to make judgements about expert advice or evidence being offered in another. So the chemist uses their knowledge about modelling, or statistics, or the way the peer review and academic systems of grant-giving, appointments, prizes and publications work, to inform their judgement about advice from a physicist.

The craft skill of enabling scientific advice best to inform policy often concerns the design and operation of the system in which such judgements are made. A highly cross-disciplinary piece of advice requires a highly cross-disciplinary review panel. The disciplinary experiences will never match exactly. Someone, somewhere, has to combine the different insights.

The challenges of effective peer review of cross-disciplinary work are exactly what hold back research in cross-disciplinary fields and make academics very nervous. But for policy it has to be done, so the craft skill is about shaping the panel, it size, representations, how to triangulate who should be on it, and who should chair it and how. In the end, for most policy-making, where neither security nor commerce make the issues confidential, the bedrock of the quality and trustworthiness of the evidence is full transparency. Those taking part, and those using it, know that if anyone wants to interrogate and challenge how the advice was arrived at, they can.

In practice, the broker uses both forms of meta-expertise; making judgements about expertise, but also about people. The policy maker should always be informed by the best science, but getting that science may not always mean putting them in the room directly with the best scientist.

THE CRAFT SKILL OF CONSTRUCTING CONVERSATIONS

What people think about science is often not the same as what scientists or politicians think they think, especially if the people are put in control of the discussion. Structured public dialogue typically involves selected groups of people with mixed socio-economic and educational backgrounds meeting with scientists, in different localities, and expertly prepared for and facilitated throughout. Done well, it is time consuming, expensive and invaluable.

There are three principal modes of operation for public dialogue[1] broadly concerned with improving communication, creating new knowledge, and co-creating new ideas. The first mode gathers social intelligence, usually for the purposes of more effective broadcast communication. It allows scientists better to understand what questions publics might want answered, rather than what knowledge the scientist thinks they ought to have. For example, scientists at the Royal Society working on providing accessible information about genetically modified crops were initially surprised to see that some people wanted to know if there were genes in food. It typically helps the scientists to respond if they are also aware of the concept of bounded rationality: people are not stupid, but the pattern-spotting human brain will make links between scientific concepts half heard or half seen that may initially, to the scientist, seem odd.

---

[1] Again, the terms are used differently by different users. Here, public dialogue refers to structured engagement in which experts and publics work together. Engagement covers a wider range of events, such as panels, debates and other forms of communication with elements of interaction.

The second mode creates or tests knowledge about what people think now and what they think when given a chance to discuss the science. Surveys can establish superficial findings about the former, but dialogue enables people, scientists and policy makers to explore questions and avenues of thought, potential opportunities and risks. This is not about influencing those who take part, nor usually about creating scalable findings, but provides proof of concept and illustration as to what people might or might not think, say or do in future debates and circumstances. It helps particularly where public debate risks becoming locked in to a particular view such as that, for example, "the public" might always want a human in the loop when an AI system is taking decisions. Dialogue shows that is not the case: given the opportunity to discuss the matter, public judgements are much more context specific.

When it comes to public engagement with emerging science and technology, successive dialogues have generated generalizable findings that are still not sufficiently part of mainstream policy or scientific thinking, let alone public debate (Doubleday & Teubner, 2012). They show that people engage with the new technology most commonly from a position of cautious optimism, in which final judgements are highly context specific. Participants are typically concerned about who is developing the technology and for what purpose; about the distribution of benefits and risks, and about issues such as access and fairness. Overall, they do not take positions for or against broad technologies, but give nuanced responses depending on the application. In the Royal Society's dialogue on Machine Learning, for example, the discussions operated both at the levels of interest in the potentially very broad impacts on humanity of future uses of AI and simultaneously at the level of highly context-specific perceptions and judgements about risk and value in different potential applications such as those in health, finance or education.

In some ways, public dialogue of this structured kind exposes the fact that, when people vote they don't have the opportunity to engage in technical discussion and that, if they do, they may think differently about the issues. But that knowledge, if developed through dialogue, isn't sufficient to legitimise a public decision that takes it in to account. For example, the use of citizens' juries for debates in which new scientific knowledge plays a major part has not yet become widespread and it is generally difficult to link insights from dialogue to wider democratic processes.

In practice, decision-makers use this kind of dialogue to challenge public and private assumptions about the inevitability of "what people think" about something and, if necessary, to determine how to frame the more formal and widespread consultation. So far the most robust sustained forms of engagement have been through structural mechanisms such as the Human Fertilisation and Embryology Authority.

The third mode of operation is to co-create new possibilities for science and for policy more directly. Researchers who take part often describe how the conversations have prompted new ideas for their future research questions. This aspect of public dialogue overlaps with futures work and with many examples of national and local innovative policy-making.

The creation of the Intergovernmental Panel for Climate Change established a system for developing and distilling evidence on matters of science relevant to policy that evolve over decades. However the timescales for the assessment of the implications of emerging technologies such as new genome editing techniques, or of new forms of data sciences appear to be shorter. They require new forms of expert and public deliberation, policy-making and regulation. Importantly, new ways forward are being crafted painstakingly by mixtures of practice and theory from many disciplines and in many areas of policy (see, for example, Burrall, 2018; Jasanoff & Hurlbut, 2018; Mulgan, 2017).

Opinion-Forming

It can be very difficult for a scientist fully to internalise the reality that information on its own doesn't change people's minds. Increasingly large volumes of practice and theory show it. Even more worryingly for some scientists, there are clear arguments that under many circumstances providing more information increases, rather than decreases, the polarisation between groups. Two people initially separated by a small difference in views on, say, whether humans are contributing to climate change, grow further and further apart as each preferentially notices and rationalises evidence that supports the direction of their initial view, and reinforces the social networks and other habits that reinforce it (see, for example, De Meyer, 2017).

It is also sometimes hard both for scientists and policymakers to come to terms with what appears to be the irreducible group of people who will not be persuaded of some scientific finding, regardless of the degree of scientific or other consensus about the evidence. Work on conspiracy theorists helps to make this position understandable, by showing how, for example, being part of a small group brings social benefits, or how it may be more comfortable to believe that someone is cotrolling a situation, rather than there being no guiding agency, even if their purposes are malign.

The links between information, opinions, discussion and decisions are increasingly being studied. For example, interesting experiments showed that under certain conditions multiple smaller groups were more accurate in estimating the correct answer to a factual question than the same number of people in a larger group. Such experiments can also begin to explore the ways groups negotiate towards a position on issues in which values as well as facts are at stake (Sigman & Ariely, 2017).

Who Sets the Questions?

It is helpful to put discussion of what publics make of the science in the context of the issue of who created the science in the first place, and for what purpose. Discussion of public engagement with science can often appear to be about downstream questions that take the direction of the science as given. In practice the issue of what research questions get asked and answered is much more iterative and connected. The challenge is more to ensure connections between potential new knowledge and potential users, than to align questions and capacity directly to answer them. Spaces for such conversations are rare.

There is much literature on this topic but, for the UK research landscape, it is worth observing three things. The first is that synthetic historical surveys show that most science has come from problem-solving (Agar, 2012; Edgerton, 2006). Even quantum theory owes its origin to problems identified by German industry. A second is that around two thirds of spend is in business, and so directly related to questions defined by the markets. A third is that UK discourse about science funding introduced the notion of curiosity-driven science in the 1980s, at least partly to balance the tendency for governments to seek to impose top-down agendas. This followed a policy decision to separate funding for government departments to conduct or contract for science directly, from funding through research councils. Since that time, direct spend by government departments outside health and defence has significantly reduced.

Finally, new ways of enabling discussions between public, government, private and philanthropic funders are emerging across the UK government. The Haldane principle, broadly expressed, seeks to ensure that whereas wide discussions about what scientific knowledge society might seek or wish to avoid are essential, judgements about what constitutes robust scientific knowledge, and which scientists are most able to generate scientifically valuable new knowledge, are best made by scientists themselves.

REFERENCES

Agar, J. (2012). *Science in the twentieth century and beyond.* Policy Press.

British Science Association. (2015). *Public attitudes to science survey.* British Science Association, Ipsos Mori, Department for Busines, Innovation and Skills.

Burrall, S. (2018). Rethink public engagement for gene editting. *Nature, 555,* 438–439.

Collins, H. (2014). *Are we all scientific experts now?* Wiley.

De Meyer, K. (2017, January 4). *Brexit, Trump and "post-truth": The science of how we become entrenched in our views.* The Conversation, UK. Retrieved May 7, 2018, from https://theconversation.com

Doubleday, R., & Teubner, R. (2012). *Public dialogue review.* Research Councils UK; Centre for Science and Policy; Involve; ScienceWise Expert Resource Centre.

Edgerton, D. (2006). *Welfare state: Britain, 1920–1970.* Cambridge University Press.

Jasanoff, S. (1990). *The fifth branch: Science advisers as policymakers.* Harvard University Press.

Jasanoff, S., & Hurlbut, J. B. (2018). A global observatory for gene editting. *Nature, 555,* 435–437.

Lentsch, J., & Weingart, P. (2009). *Scientific advice to policy making: International comparison.* Barbara Budrich Publishers.

Mulgan, G. (2017). *Big mind: How collective intelligence can change our world.* Princeton University Press.

Neuberger, D. (2016). Stop needless dispute of science in the courts. *Nature, 531,* 9.

Owens, S. (2015). *Knowledge, policy and expertise.* Oxford University Press.

Sigman, M., & Ariely, D. (2017, April). How can groups make good decisions. *TED Studio.* Retrieved April 22, 2018, from https://www.ted.com/talks/mariano_sigman_and_dan_ariely_how_can_groups_make_good_decisions/up-next

The Royal Society & The Royal Academy of Engineering. (2012). *Shale gas extraction in the UK: A review of hudraulic fracturing.* The Royal Society & The Royal Academy of Engineering. Retrieved April 22, 2018, from https://royalsociety.org

How to Ensure That When a Minister Meets a Nobel Laureate They Both Have a Great Encounter

THE THREE-BODY PROBLEM

This chapter considers one particular moment: an imaginary encounter between a Minister and a Nobel Laureate. Embodiment, in a very human sense, is at the heart of the moment. Drawing on celestial mechanics for the metaphor, there is what physicists call a three-body problem: in which the Minister, the Nobel Laureate and, in a very different way, the broker, all play parts.

The Minister and the Nobel Laureate each have the positions they hold due to the existence of a long-standing system of human organisation: one democracy, one science. Both will have been delivered up to the top of their systems through tough competition. Their objectives, what they know, how they think and what enables and constrains them, are shaped by that system.

To simplify greatly: the politician is answerable through elections to the citizens, and viewed through the windows of reportage and social media. The scientist is accountable by peer review and challenge largely from within their discipline and viewed through the lenses of published literature. When in the room together at some particular moment, both systems require the human to combine those responsibilities with their personal judgement and instincts. And, if they care about science and government at all, to take the empathetic step of knowing that what they do will have to be explained in the other's system too.

© The Author(s) 2019
C. Craig, *How Does Government Listen to Scientists?*,
https://doi.org/10.1007/978-3-319-96086-9_5

The politician's position appears more fragile than the scientist's and in some ways the greater freedom is with the scientist. Government generally wants to learn from science, even if doesn't always act on the knowledge. The politician will typically have to justify what they do when they do or don't take account of the evidence: but they can rarely blame or criticise the science or the scientist. Relative levels of public trust in the two groups remain consistently higher for scientists.

The scientist, on the other hand, gains with some audiences if their advice appears to lead directly to well-informed outcomes, and with others if they are seen as having boldly spoken unpalatable truth to power even if it is not acted upon immediately. Indeed, the occasional academic can seem to take the damaging view that only knowledge that is *not* acted upon could have been fully intellectually independent in the first place.

But that asymmetry is truer in private than in public. The old presumption that only second rate scientists become "communicators" is largely dead, but scientists still risk their reputations with some of their colleagues simply by engaging in policy debates at all. It is fundamental to the provision of scientific evidence that for a scientist to make public statements about science in the policy context can be extremely risky for them. Scientific statements tend to get more scientifically robust the narrower, more specific, more qualified and more technically and therefore accurately expressed they are. As they move further along these dimensions they get less publicly intelligible and further from being accessible and informative to the policy maker. The broader, more general, less qualified and less technically expressed statement is more intelligible to the Minister but inevitably opens the scientist to criticism from their peers for inaccuracy.

The scientist faces a further risk if they speak—as public policy discussions tend to compel them to do—on topics outside their very immediate field of research or scholarship. This criticism of speaking beyond your academic domain is a stinging one inside academic discussions, where even asking an apparently poorly-framed question can cause embarrassment.

Another asymmetry is their likely confidence about understanding each other's systems. Political life and politicians are subject to extensive public scrutiny and debate. The scientist may be looking at the politician through a window made blurry by the selectiveness of media coverage and gossip, but they will have been able to vote since they were 18 and will often think they have some grasp of the current top-level political situation. By contrast, except in rare cases, the politician will know little about how research works, the incentives and pressures on the Nobel Laureate, and may have given up all forms of formal science at age 16.

One symmetry, at the very top of the professions, is that they are most likely to be men. For example in the UK outside the health sector, there have been few Chief Scientific Advisers who are women and no Government Chief Scientific Advisers (GCSAs). That will change, and there are many cases for celebration of diversity such as the fact that, for at least a few years, the photograph of the Prime Minister meeting the President of the Royal Society and Nobel Laureate would be the photograph of a woman meeting a UK-US citizen born in India.

Then the third body, not always actually in the room, is that of the broker. This is not the Honest Broker of Pielke's roles for science advisers, but the intermediary. For the Minister the broker is likely to be a civil servant, but such brokers exist in the public, private and not-for-profit sectors. This person, with others like them, has been trying, possibly for months or even years, to get the arrangements right. The expectations on both sides ahead of the meeting, the framing, where and when the meeting happens, who is present, how they are briefed, whether a brief matters, what happens immediately before or after, and everything that might enable the two to make the best possible use of their time, to find common ground for well-founded discussion or decision-making.

The brokering takes a particular importance because of the fragility of the moment. Under most circumstances, neither of the main protagonists is professionally obliged to build a continuing relationship with the other nor to encourage others of their tribe to engage across the interface. But if the encounter and its outcomes go well, more will ensue.

CRAFT AND CURATION

There is a rarely acknowledged element of craft skill here. The encounter may be affected by where the participants meet, sit or stand, the provision of food and drink or tiny adjustments to their physical needs of sight, hearing or movement. In adrenaline filled moments of crisis or opportunity none of this matters. But for the usually slow and careful process of building relationships, enabling the growth of respect and empathy, of drawing ideas across sectors or disciplines, such details can make a surprising difference.

Senior figures have objected to orchestrated events where there were too many, or not enough, young people in the room. They have turned down events because the location didn't also provide suitable photo opportunities. Participants nearly collapsed because austerity measures meant there was no water available in a hot meeting room. Encounters failed because they were held early in the morning when the senior person was invariable less tolerant to slow thinkers; dinner discussions failed because there was too much alcohol, while other encounters would have gone better with some. Both sides can laugh at formal facilitation, or demand it, depending on their experience and preference. Senior civil servants don't always reply or read emails and given the volumes they get they couldn't function if they did. Partly for the same reason, internally they rarely use extensive salutations. In the culture of the top of the public sector extensive greetings and sign off may be perceived as unnecessarily wasting the time both of sender and recipient. However academics, certainly with those they don't know well, tend to spend more time on salutations as part of relationship building and may misinterpret their omission.

Language also matters. Every senior person has words that are important to them or which they reject and, like mirroring body language, choosing to use preferred words can help the conversation. Their choice of words may reflect quite deliberate signalling: the knowledge that a well-chosen clear but slightly unusual word or phrase, repeated often, carries with it influence within and beyond the immediate organisation. It may be more of a habit of dislike, or a view that certain language indicates intellectual sloppiness: so including it in communicating with them wastes their cognitive energy. Depending on the setting they may not ask for clarification when they don't understand a word or argument.

For larger events the required skill may be better described as a form of curation. The broker working with scientists, in particular, needs to overcome the implicit assumption and modes of working in much of research which naturally respect the strength of ideas over the humans expressing them. There is a splendid integrity in this: to be persuasive in the moment is not to have your scientific ideas "win" over time. Sometimes the sheer lack of polish of itself is seen as proof of integrity, but the converse is not true. The best evidence, poorly expressed, may miss the opportunity to inform a Minister's mind. The public policy roadshow moves on, and it can be a long time until the next receptive moment.

This curation of events is also integral to enabling the effective working across disciplines which is itself essential to the evolution of ideas and provision of advice. Working across disciplines requires the creation of an environment in which participants are curious about each other, open-minded, courteous, and confident. The curation often requires a slowing down of the project to create a firm basis for the cross-disciplinary discussion. Whether it is a single workshop or a major project, collaboration needs to start with the expression of what is known and relevant from each discipline or sector, in language that all the others can grasp. That then enables a collective discussion about the challenge at hand. It is a mistake to leap straight into that discussion and discover minutes or months later that there were fundamental areas of ignorance or misconception. As discussed in Chap. 2, the biggest misunderstandings may come about when people think they mean the same thing because they are using the same word, but the mental models and associations behind it are different.

RHYTHMS

Syncopation can be hard too. The caricature is that science is too slow: that a researcher's response to being told of a policy question is to ask for funding and several years' time in order to come back with the answer. As one Chief Scientific Adviser puts it "many quite complex policy decisions are taken in days, weeks or at most a few months, the time it takes a competent PhD student to begin an introductory chapter" (Whitty, 2015).

In truth, science is also in some ways too fast, or at least too fragmented. Its ragged frontier means there are continuous new findings on controversial issues from climate change to neonicotinoid pesticides. The directional conclusions of science move very slowly, which is helpful to building public consensus for policy choices. But the continual rapid fire of specific new findings, and the way they are promoted by individual scientists and in the media, can destabilise public assumptions about the degree of scientific consensus and create space for dispute on the fundamentals in areas such as climate change, where there should be none.

UK policy-making has multiple rhythms: from the decision that needs to be taken today to the 6–9 months legislative period between the politi cal Party Conferences and the summer Recess, the annual cycle of Budgets and the typically 3–5 year cycle of major Spending Reviews. Then there are, of course, the deeper currents that create substantial policy changes over decades.

There are different ways of science being involved in each of those. Effective advisers and skilful policy-makers sometimes build trust and relationships by giving and using the readily available scientific evidence for short term issues (such as during emergencies or for relatively narrow questions) and then working together over longer periods on broader questions and more difficult agendas (such as the impacts of climate science or the evolution of wholly new concepts such as human wellbeing).

SYSTEMS AND TRIBES

There is a remedy to the challenges of the personal risks to individual scientists of providing science advice, and of the risks to the policy maker from the mismatch between single discipline knowledge and the breadth of evidence required by policy decisions. It is to provide the evidence through institutions (Oreskes & Conway, 2010).

Each national political and scientific ecosystem has a different and continually evolving set of institutions. The UK has had a GCSA in one form or another since the Second World War. The postholder has typically been supported by their own set of brokers, currently forming the Government Office for Science, capable of delivering reports on timescales from hours to more than a year. At European level, the EU's Joint Research Centre is now described as the European Commission's science and knowledge service. With 3000 people and a budget of over 300 m Euros, its Knowledge Management Centres are intended to provide continuous access to the latest data, knowledge and expertise in areas from modelling to disaster relief and migration.

National academies and international networks of national academies provide potentially powerful vehicles for authoritative inter-disciplinary statements. They bring together senior researchers who can then draw both on their disciplinary or domain knowledge and their knowledge of the wider scientific and scholarly systems. This enables them to make collective judgements about both narrow and broad areas of knowledge. Given the relatively small levels of resource and researchers' time available, their key challenge is the one of selecting and framing the most valuable issues. Working exclusively too close to today's debates risks missing the moment or failing to spot future and profound issues. Exploring issues too far away from today's concerns risks becoming irrelevant to policy makers and the public and so losing influence in the long term.

Increasing in number and range in the UK are the policy-facing institutes in universities, such as the Institute for Policy Research at Bath, the Cabot Institute at Bristol or the Policy Institute at King's College London. The growth of these institutes has in part been driven by the behavioural and cultural shift which reflects the inclusion of impact as a measure of research value and as a factor affecting funding. They differ in scope and mode of operation from a small policy-facing communication office to cross-disciplinary research institutions embedded in research and teaching within the university. Using information from such different sources itself requires the broker to exercise the craft skill of knowing who to listen to, how to combine, and how best and proportionately to quality assure.

In her study of the Royal Commission on Environmental Pollution Susan Owens points up powerfully how it was the sustained engagement between scientists and policy makers that built relationships and insights across the tribal boundaries and meant the selection of issues, their investigation and the findings were better informed and had more impact (Owens, 2015). Drawing on case studies of areas from the introduction of carbon budgets and the banning of smoking in public, to the introduction of the minimum wage, the Institute for Government reviewed the conditions under which the UK saw substantial policy change at least in part due to new scientific evidence. These studies also suggested that one of the conditions for success was the existence of sustained professional relationships by individuals in networks that crossed between academia and government (Rutter, Sims, & Marshall, 2012).

This longer term and important challenge of changing the cultures of research and policy so that more individuals are comfortable working across the boundaries is part of a wider set of currents informing both systems. For research culture, it can helpfully be part of a trend towards diversity in all its dimensions and towards enabling easier interchange between those in academia and those working in the business or the public sector at every career stage. In addition to the emergence of "braided" careers in which researchers might move between sectors or hold posts in two sectors at the same time, there are many opportunities to sample and engage. For scientists in many universities and research institutes they include PhD and undergraduate training, Early Career Fellowships or placements, and mid and senior career opportunities to engage in public events, academies' projects, or government advisory groups.

For policy-makers looking to increase their exposure to research, the opportunities include Parliamentary and other pairing schemes, and increasing numbers of continuing professional development opportunities. One internationally exceptional example of the craft skill of changing cultures is the Centre for Science and Policy, a project with a simple model at its core. In this, policy makers, typically fast track civil servants, bid for the opportunity to spend five days at Cambridge University meeting academics with whom they could explore one or more questions that had previously identified as important to their future work. A clear central proposition, careful selection of the people on both sides, intensive use of the time (right down to the management of the travel time between Cambridge locations) and constant attention to feedback and detail, has helped the programme build extensive relationships between members of its target community.

TRUST, POST TRUTH AND WHERE NEXT

I once heard a Minister say that scientists are cold, while politics is hot. However, as Chap. 1 began to explore, in some senses this couldn't be less true. Scientists are typically driven, emotional, imaginative, deeply competitive (with themselves and with others) people. They often thrive in large teams, and compete and collaborate with bewildering complexity. It is the scientific system that imposes a form of integrity that transcends the individuals and ensures that science continually attempts to come back to mindfulness of how the world is, rather than how a particular scientist might imagine, hypothesise or wish it to be.

In the 1950s when CP Snow wrote about two cultures he spoke of the arts and the sciences (Snow, 1959). Then and for much of the intervening time, policy-makers have typically seemed or seemed to see themselves as closer to the arts. Today's "post-truth" debates cut directly across this binary divide. Rigour, logic, creativity and imagination are exclusive neither to the arts and humanities nor to the sciences. The twenty-first century's two cultures might be better expressed as rationality and sentiment. That is, if there are only two: perhaps the point is that public debate better accepts the plurality of world views but doesn't know how to describe them or to describe the process of simply getting things done—paying taxes, stopping violence, feeding people and keeping the lights on—in ways that accommodate them.

In the UK public trust in scientists has not gone down in the last few decades. Presumably that is trust in scientists to be scientists, whatever that means. So whatever the post-truth world is, it still appears to have a place for science. Similarly, the popular presumption that expertise is now tarnished is clearly a worrying one, but not all expertise and not all of the time. Perhaps we need more to emphasise human failings and the distinction between science and scientists. It's a "memento mori" to the Nobel Laureate that there is evidence that in some circumstances experts are more likely to be wrong, the more confident they are of being correct. Individual expertise is not the same as scientific knowledge.

The reality is that scientists and policymakers—as human beings—rely on both emotion and cognition. A great encounter allows them to engage through both.

References

Oreskes, N., & Conway, E. M. (2010). *Merchants of doubt*. Bloomsbury Press.

Owens, S. (2015). *Knowledge, policy and expertise*. Oxford University Press.

Rutter, J., Sims, S., & Marshall, E. (2012). *The "S" factors: Lessons from IFG's policy success reunions*. Institute for Government. Retrieved April 29, 2018, from https://www.instituteforgovernment.org.uk

Snow, C. P. (1959). *The two cultures and the scientific revolution*. Cambridge University Press.

Whitty, C. J. (2015). What makes an academic paper useful for health policy? *BMC Medicine, 13*, 301.

Index[1]

[1] Note: Page numbers followed by 'n' refer to notes.

© The Author(s) 2019
C. Craig, *How Does Government Listen to Scientists?*,
https://doi.org/10.1007/978-3-319-96086-9

69